Learn

Eureka Math™

Grade 1
Modules 2 & 3

Published by Great Minds®.

Printed in the U.S.A.
This book may be purchased from the publisher at eureka-math.org.
10 9 8 7 6 5 4 3 2 1

v1.0 PAH

ISBN 978-1-64054-051-4

G1-M2-M3-L-05.2018

Learn • Practice • Succeed

Eureka Math™ student materials for *A Story of Units*® (K–5) are available in the *Learn, Practice, Succeed* trio. This series supports differentiation and remediation while keeping student materials organized and accessible. Educators will find that the *Learn, Practice,* and *Succeed* series also offers coherent—and therefore, more effective—resources for Response to Intervention (RTI), extra practice, and summer learning.

Learn

Eureka Math Learn serves as a student's in-class companion where they show their thinking, share what they know, and watch their knowledge build every day. *Learn* assembles the daily classwork—Application Problems, Exit Tickets, Problem Sets, templates—in an easily stored and navigated volume.

Practice

Each *Eureka Math* lesson begins with a series of energetic, joyous fluency activities, including those found in *Eureka Math Practice*. Students who are fluent in their math facts can master more material more deeply. With *Practice*, students build competence in newly acquired skills and reinforce previous learning in preparation for the next lesson.

Together, *Learn* and *Practice* provide all the print materials students will use for their core math instruction.

Succeed

Eureka Math Succeed enables students to work individually toward mastery. These additional problem sets align lesson by lesson with classroom instruction, making them ideal for use as homework or extra practice. Each problem set is accompanied by a Homework Helper, a set of worked examples that illustrate how to solve similar problems.

Teachers and tutors can use *Succeed* books from prior grade levels as curriculum-consistent tools for filling gaps in foundational knowledge. Students will thrive and progress more quickly as familiar models facilitate connections to their current grade-level content.

Students, families, and educators:

Thank you for being part of the *Eureka Math*™ community, where we celebrate the joy, wonder, and thrill of mathematics.

In the *Eureka Math* classroom, new learning is activated through rich experiences and dialogue. The *Learn* book puts in each student's hands the prompts and problem sequences they need to express and consolidate their learning in class.

What is in the Learn *book?*

Application Problems: Problem solving in a real-world context is a daily part of *Eureka Math*. Students build confidence and perseverance as they apply their knowledge in new and varied situations. The curriculum encourages students to use the RDW process—Read the problem, Draw to make sense of the problem, and Write an equation and a solution. Teachers facilitate as students share their work and explain their solution strategies to one another.

Problem Sets: A carefully sequenced Problem Set provides an in-class opportunity for independent work, with multiple entry points for differentiation. Teachers can use the Preparation and Customization process to select "Must Do" problems for each student. Some students will complete more problems than others; what is important is that all students have a 10-minute period to immediately exercise what they've learned, with light support from their teacher.

Students bring the Problem Set with them to the culminating point of each lesson: the Student Debrief. Here, students reflect with their peers and their teacher, articulating and consolidating what they wondered, noticed, and learned that day.

Exit Tickets: Students show their teacher what they know through their work on the daily Exit Ticket. This check for understanding provides the teacher with valuable real-time evidence of the efficacy of that day's instruction, giving critical insight into where to focus next.

Templates: From time to time, the Application Problem, Problem Set, or other classroom activity requires that students have their own copy of a picture, reusable model, or data set. Each of these templates is provided with the first lesson that requires it.

Where can I learn more about Eureka Math *resources?*

The Great Minds® team is committed to supporting students, families, and educators with an ever-growing library of resources, available at eureka-math.org. The website also offers inspiring stories of success in the *Eureka Math* community. Share your insights and accomplishments with fellow users by becoming a *Eureka Math* Champion.

Best wishes for a year filled with aha moments!

Jill Diniz

Jill Diniz
Director of Mathematics
Great Minds

The Read–Draw–Write Process

The *Eureka Math* curriculum supports students as they problem-solve by using a simple, repeatable process introduced by the teacher. The Read–Draw–Write (RDW) process calls for students to

1. Read the problem.
2. Draw and label.
3. Write an equation.
4. Write a word sentence (statement).

Educators are encouraged to scaffold the process by interjecting questions such as

- What do you see?
- Can you draw something?
- What conclusions can you make from your drawing?

The more students participate in reasoning through problems with this systematic, open approach, the more they internalize the thought process and apply it instinctively for years to come.

Contents

Module 2: Introduction to Place Value Through Addition and Subtraction Within 20

Topic A: Counting On or Making Ten to Solve *Result Unknown* and *Total Unknown* Problems

Topic B: Counting On or Taking from Ten to Solve *Result Unknown* and *Total Unknown* Problems

Grade 1
Module 2

Read

John, Emma, and Alice each had 10 raisins. John ate 3 raisins, Emma ate 4 raisins, and Alice ate 5 raisins. How many raisins do they each have now? Write a number bond and a number sentence for each.

Draw

Write

Name ______________________________ Date ______________

Read the math story. Make a simple math drawing with labels. Circle 10 and solve.

1. Bill went to the store. He bought 1 apple, 9 bananas, and 6 pears. How many pieces of fruit did he buy in all?

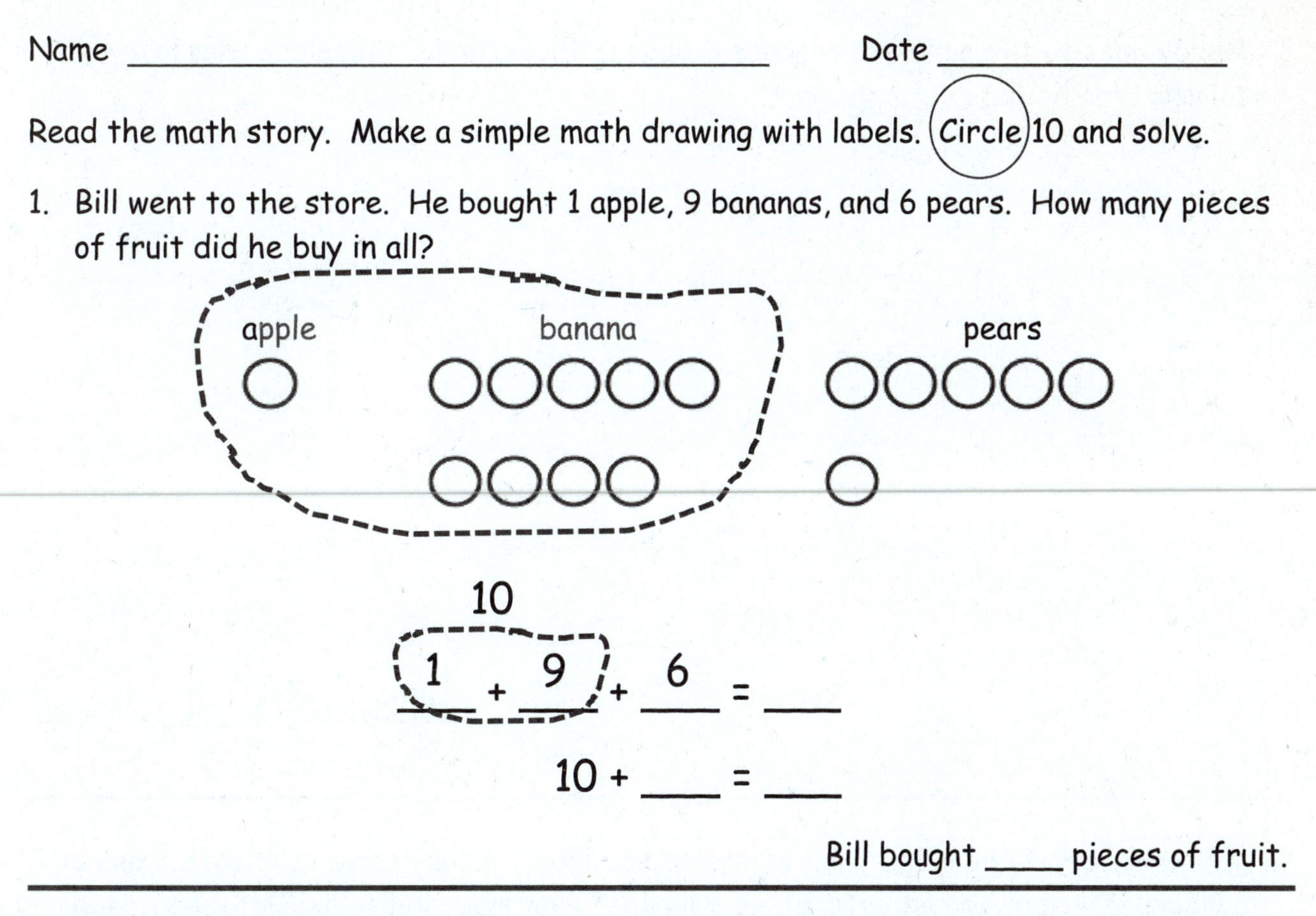

Bill bought ____ pieces of fruit.

2. Maria gets some new toys for her birthday. She gets 4 dolls, 7 balls, and 3 games. How many toys did she receive?

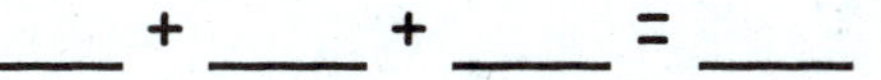

10 + ___ = ___

Maria received ____ toys.

3. Maddy goes to the pond and catches 8 bugs, 3 frogs, and 2 tadpoles. How many animals did she catch altogether?

____ + ____ + ____ = ____

10 + ____ = ____

Maddy caught ____ animals.

---

4. Molly arrived at the party first with 4 red balloons. Kenny came next with 2 green balloons. Dara came last with 6 blue balloons. How many balloons did these friends bring?

____ + ____ + ____ = ____

10 + ____ = ____

There are ____ balloons.

Name ________________________________ Date ______________

Read the math story. Make a simple math drawing with labels. Circle 10 and solve.

Toby has ice cream money. He has 2 dimes. He finds 4 more dimes in his jacket and 8 more on the table. How many dimes does Toby have?

___ + ___ + ___ = ___

10 + ___ = ___

Toby has ____ dimes.

Read

Lisa was reading a book. She read 6 pages the first night, 5 pages the next night, and 4 pages the following night. How many pages did she read?

Make a drawing to show your thinking. Write a statement to go with your work.

Extension: If she read a total of 20 pages by the fifth night, how many pages could she have read on the fourth night and the fifth night?

Draw

Write

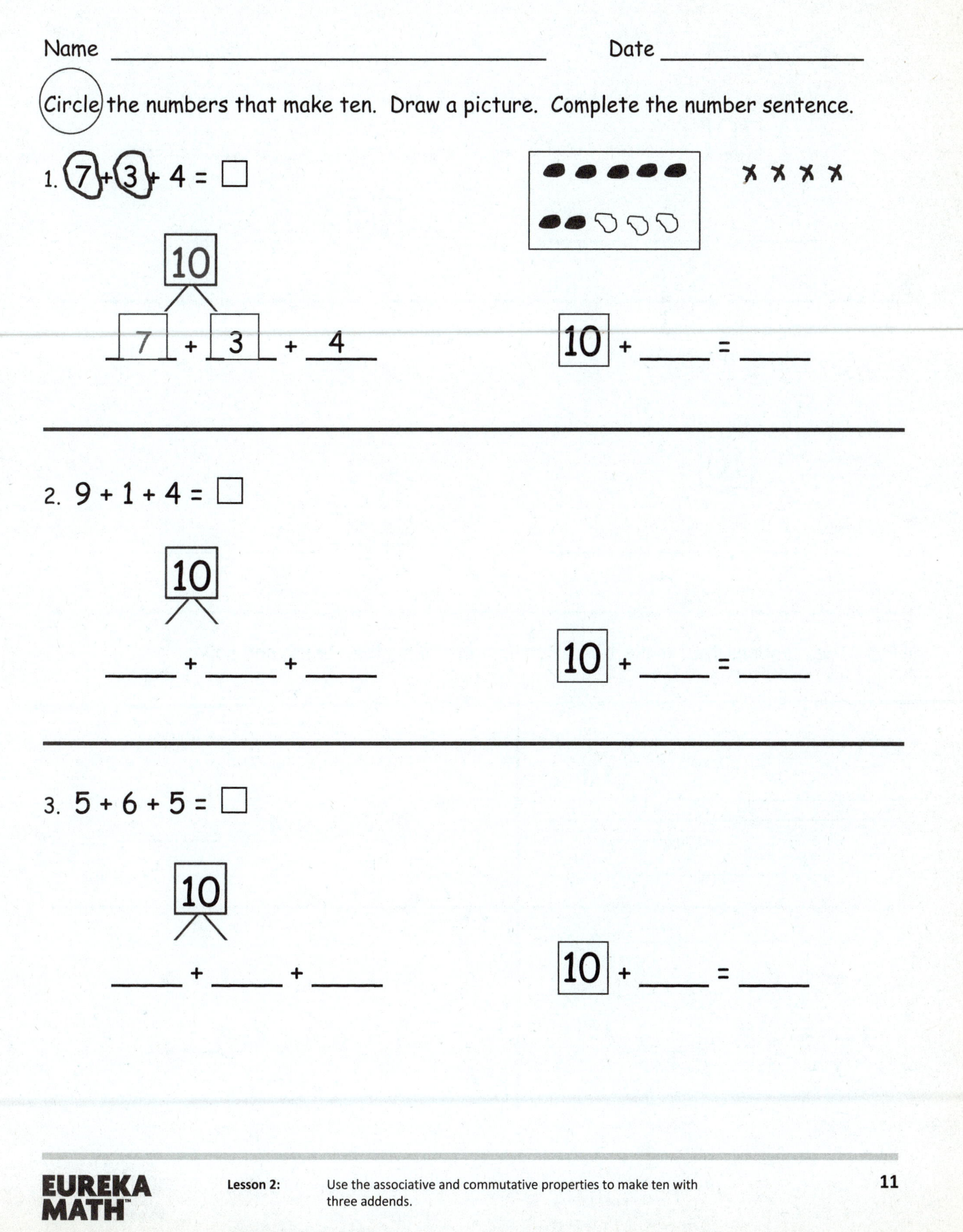

Name ______________________ Date ____________

Circle the numbers that make ten. Draw a picture. Complete the number sentence.

1. 7 + 3 + 4 = ☐

10

7 + 3 + 4

10 + ____ = ____

2. 9 + 1 + 4 = ☐

10

____ + ____ + ____

10 + ____ = ____

3. 5 + 6 + 5 = ☐

10

____ + ____ + ____

10 + ____ = ____

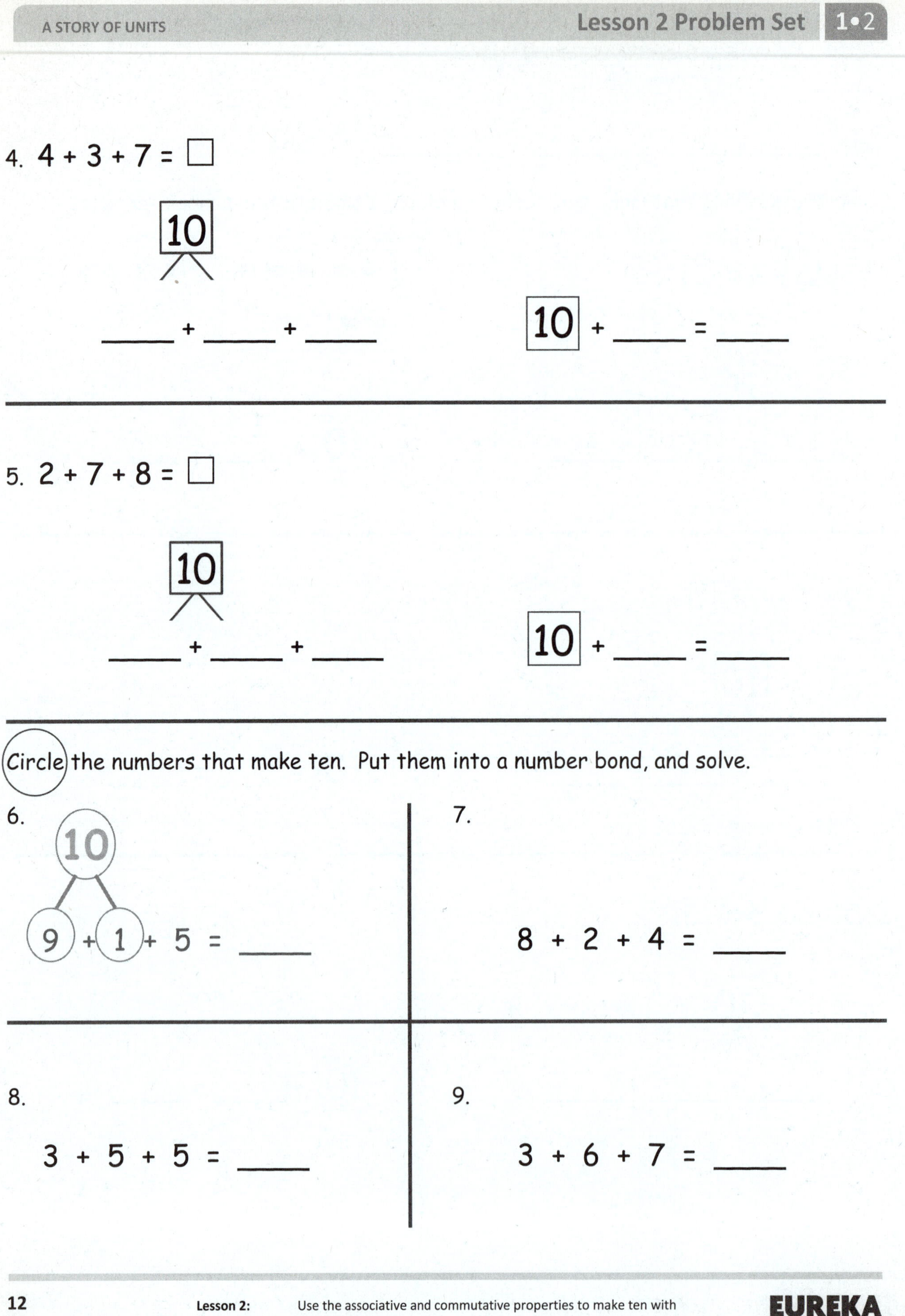
4. 4 + 3 + 7 = ☐
10
___ + ___ + ___
10 + ___ = ___
5. 2 + 7 + 8 = ☐
10
___ + ___ + ___
10 + ___ = ___
Circle the numbers that make ten. Put them into a number bond, and solve.
6.
10
9 + 1 + 5 = ___
7.
8 + 2 + 4 = ___
8.
3 + 5 + 5 = ___
9.
3 + 6 + 7 = ___

Name ______________________________ Date ______________

Circle the numbers that make ten.

Draw a picture, and complete the number sentences to solve.

a. 8 + 2 + 3 = ____

____ + ____ = ____

10 + ____ = ____

---

b. 7 + 4 + 3 = ____

____ + ____ = ____

10 + ____ = ____

Read

Tom's mother gave him 4 pennies. His father gave him 9 pennies. His sister gave him enough pennies so that he now has a total of 14. How many pennies did his sister give him? Use a drawing, a number sentence, and a statement.

Extension: How many more would he need to have 19 pennies?

Draw

Write

Name ______________________________ Date ______________

Draw and circle to show how you made ten to help you solve the problem.

1. Maria has 9 snowballs, and Tony has 6. How many snowballs do they have in all?

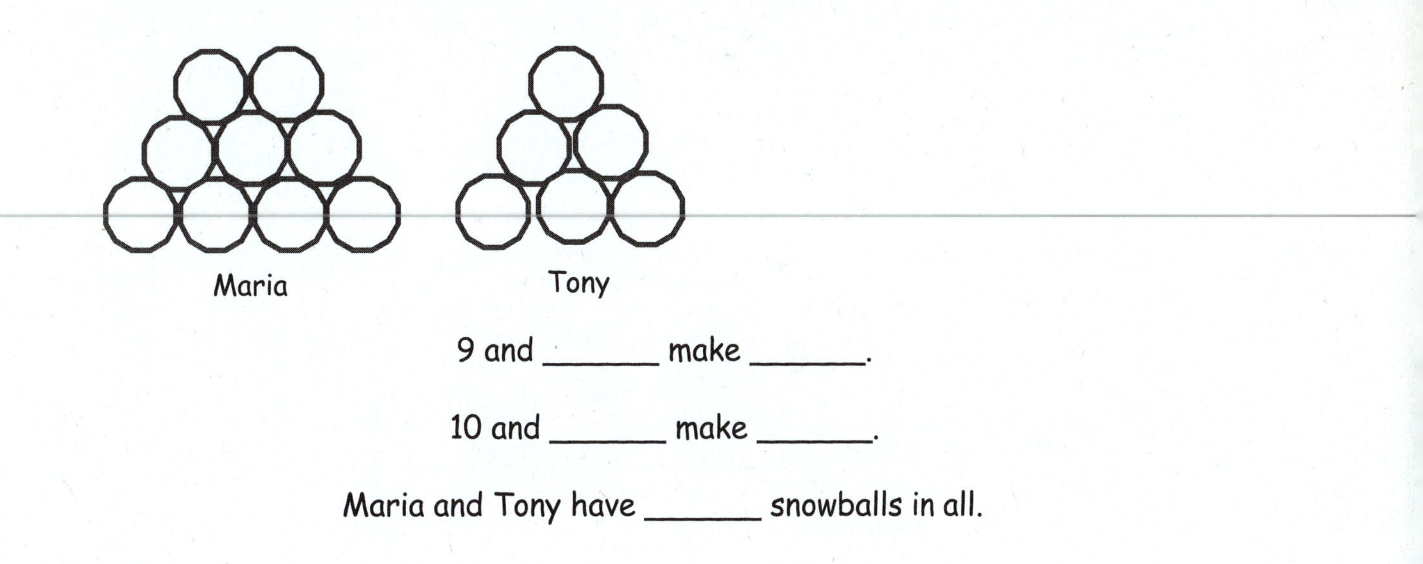

9 and _____ make _____.

10 and _____ make _____.

Maria and Tony have _____ snowballs in all.

2. Bob has 9 raisins, and Jonny has 4. How many raisins do they have altogether?

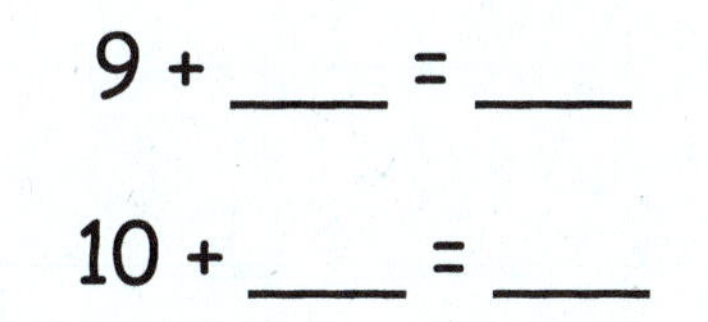

9 + ___ = ___

10 + ___ = ___

Bob and Jonny have _____ raisins altogether.

3. There are 3 chairs on the left side of the classroom and 9 on the right side. How many total chairs are in the classroom?

9 + ____ = ____

10 + ____ = ____

There are _______ total chairs.

4. There are 7 children sitting on the rug and 9 children standing. How many children are there in all?

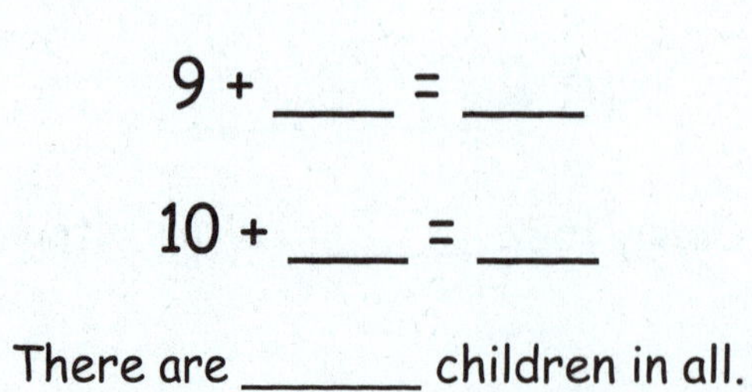

9 + ____ = ____

10 + ____ = ____

There are ______ children in all.

Name ______________________________ Date ______________

Draw and circle to show how to make ten to solve. Complete the number sentences.

Tammy has 4 books, and John has 9 books. How many books do Tammy and John have altogether?

____ + ____ = ____

____ + ____ = ____

Tammy and John have ____ books.

Read

Michael plants 9 flowers in the morning. He then plants 4 flowers in the afternoon. How many flowers did he plant by the end of the day? Make a drawing, a number bond, and a statement.

Draw

Write

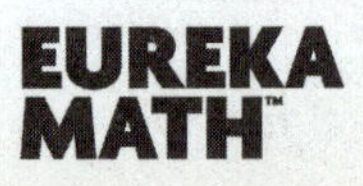

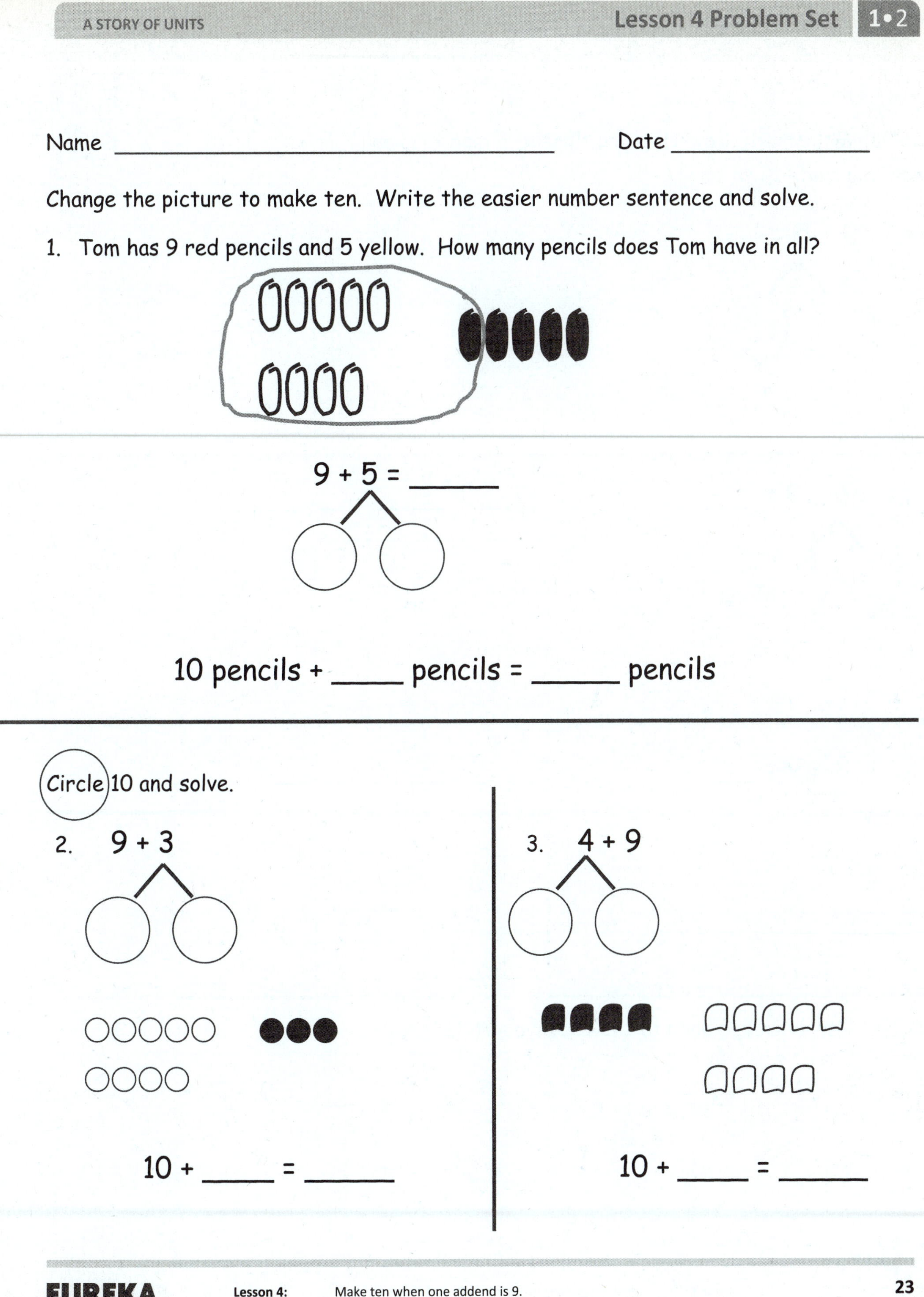

Name ______________________ Date ____________

Change the picture to make ten. Write the easier number sentence and solve.

1. Tom has 9 red pencils and 5 yellow. How many pencils does Tom have in all?

9 + 5 = _____

10 pencils + _____ pencils = _____ pencils

Circle 10 and solve.

2. 9 + 3

10 + _____ = _____

3. 4 + 9

10 + _____ = _____

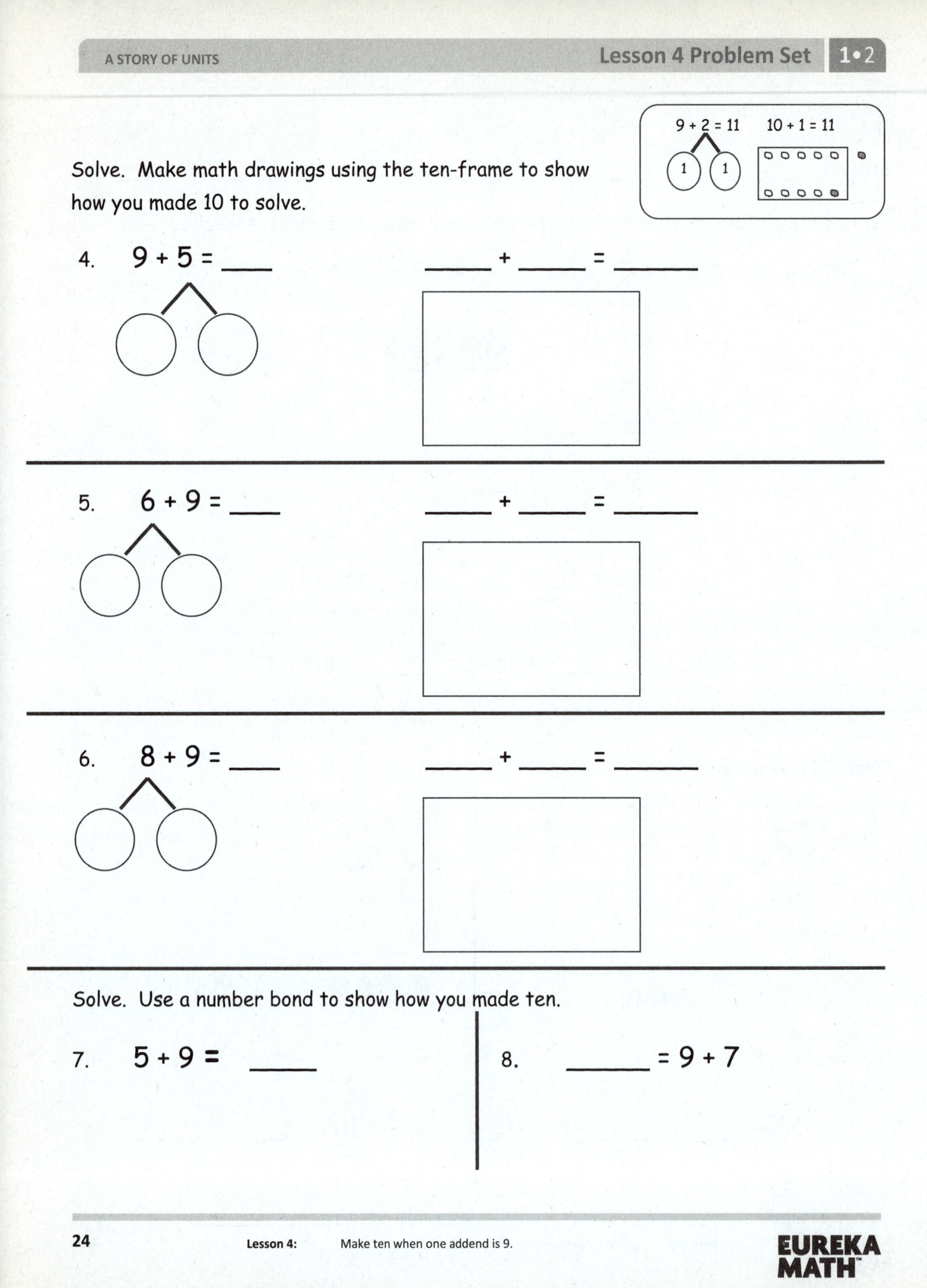

Solve. Make math drawings using the ten-frame to show how you made 10 to solve.

4. 9 + 5 = ___ ___ + ___ = ___

5. 6 + 9 = ___ ___ + ___ = ___

6. 8 + 9 = ___ ___ + ___ = ___

Solve. Use a number bond to show how you made ten.

7. 5 + 9 = ___

8. ___ = 9 + 7

Name ____________________________ Date ______________

Solve.

Make math drawings using the ten-frame to show how you made 10 to solve.

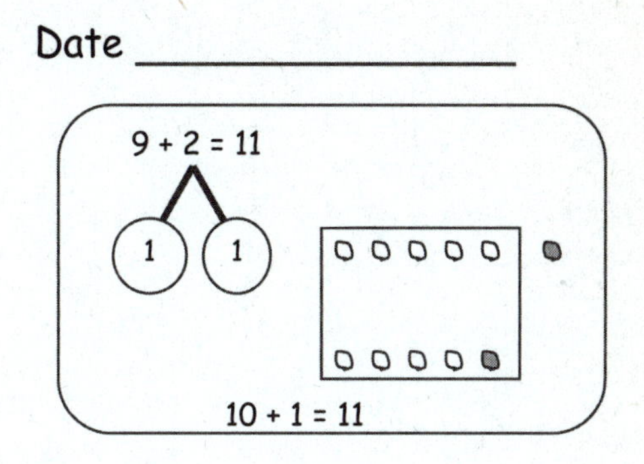

1. 6 + 9 = ____

2. ____ = 4 + 9

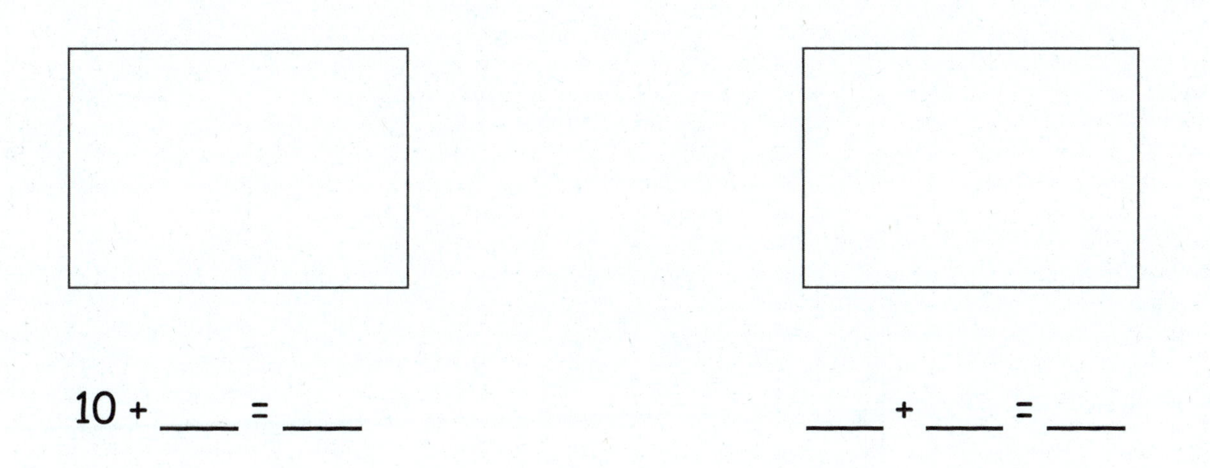

10 + ____ = ____

____ + ____ = ____

Read

There are 9 red birds and 6 blue birds in a tree. How many birds are in the tree? Use a ten-frame drawing and a number sentence. Write a number bond to match the story and a number bond to show the matching 10+ fact. Write a statement.

Draw

Write

____

____

____

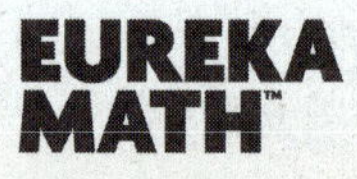

Name ______________________________ Date ______________

Make ten to solve. Use the number bond to show how you took the 1 out.

1. Sue has 9 tennis balls and 3 soccer balls. How many balls does she have?

9 + 3 = _____ 10 + ____ = ____

Sue has _____ balls.

2. 9 + 4 = _____ 10 + ____ = ____

Use number bonds to show your thinking. Write the 10+ fact.

3. 9 + 2 = _____ _____ + _____ = _____

4. 9 + 5 = _____ _____ + _____ = _____

5. 9 + 4 = _____ _____ + _____ = _____

6. 9 + 7 = ____ ____ + ____ = ____

7. 9 + ____ = ____ 10 + 7 = ____

Complete the addition sentences.

8. a. 10 + 1 = ____ (11)

b. 9 + 2 = ____ (11)

9. a. 10 + 8 = ____ (18)

b. 9 + 9 = ____ (18)

10. a. 10 + 7 = ____

b. 9 + 8 = ____

11. a. 5 + 10 = ____

b. 6 + 9 = ____

12. a. 6 + 10 = ____

b. 7 + 9 = ____

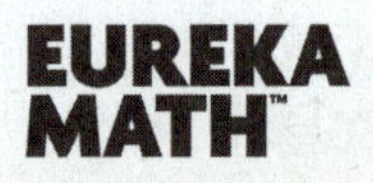

Name ________________________ Date ____________

Complete the number sentence.
Use an efficient strategy to solve the number sentences.

1. 9 + 2 = ___

2. 7 + 9 = ___

3. ___ = 9 + 5

Read

There are 6 children on the swings and 9 children playing tag. How many children are playing on the playground? Make ten to solve. Create a drawing, a number bond, and a number sentence along with your statement.

Draw

Write

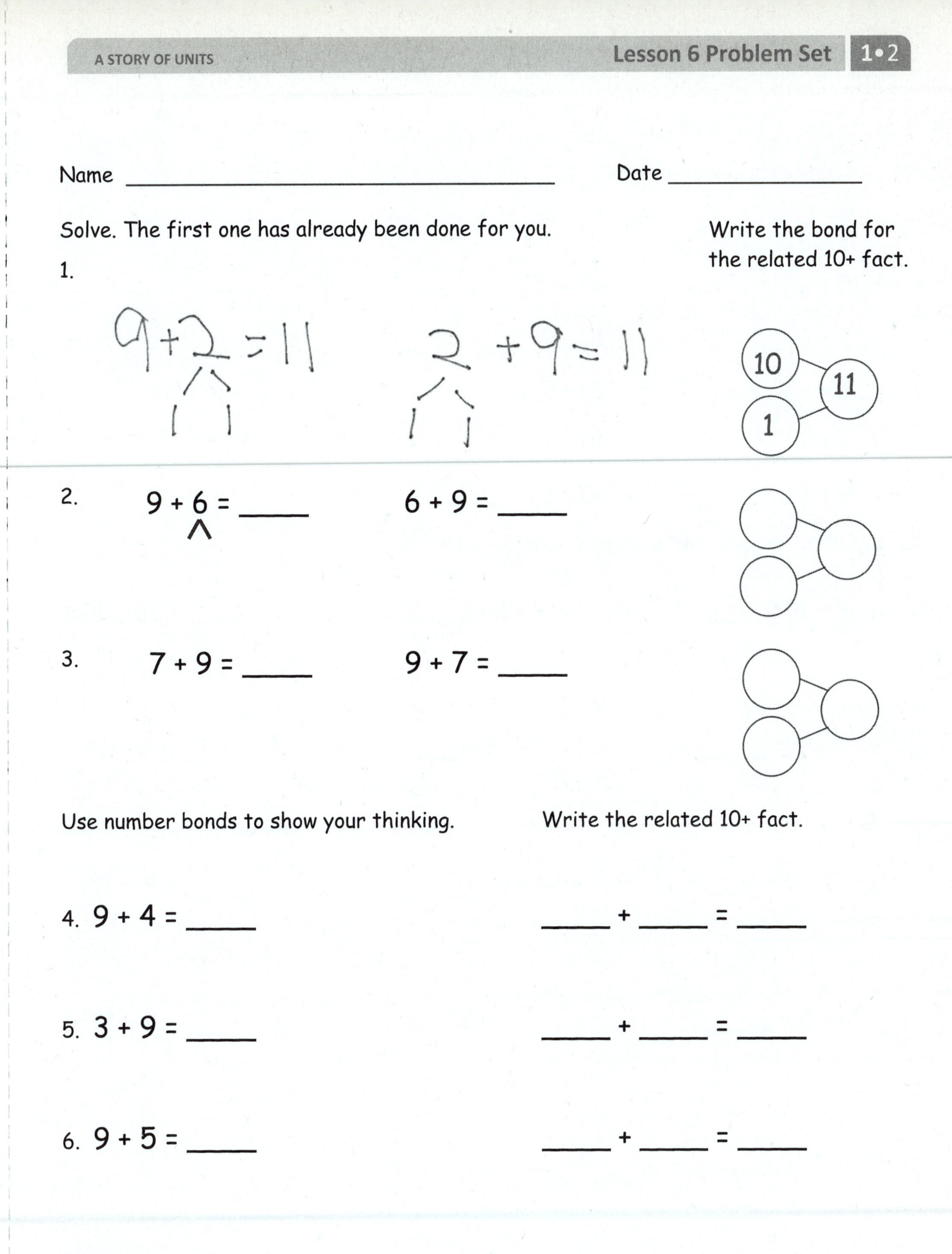

Name ______________________ Date ____________

Solve. The first one has already been done for you.

Write the bond for the related 10+ fact.

1. 9 + 2 = 11 2 + 9 = 11

2. 9 + 6 = _____ 6 + 9 = _____

3. 7 + 9 = _____ 9 + 7 = _____

Use number bonds to show your thinking.

Write the related 10+ fact.

4. 9 + 4 = _____ _____ + _____ = _____

5. 3 + 9 = _____ _____ + _____ = _____

6. 9 + 5 = _____ _____ + _____ = _____

7. Match the equal expressions.

a. 9 + 3	10 + 4
b. 5 + 9	10 + 0
c. 9 + 6	10 + 2
d. 8 + 9	10 + 5
e. 9 + 7	10 + 7
f. 9 + 1	10 + 6

8. Complete the addition sentences to make them true.

a. 2 + 10 = _____

b. 7 + 9 = _____

c. _____ + 10 = 14

d. 3 + 9 = _____

e. 3 + 10 = _____

f. _____ + 9 = 14

g. 10 + 9 = _____

h. 8 + 9 = _____

i. _____ + 7 = 17

j. 5 + 9 = _____

k. ____ + 10 = 18

l. _____ + 9 = 17

m. 6 + 10 = _____

n. _____ + 9 = 16

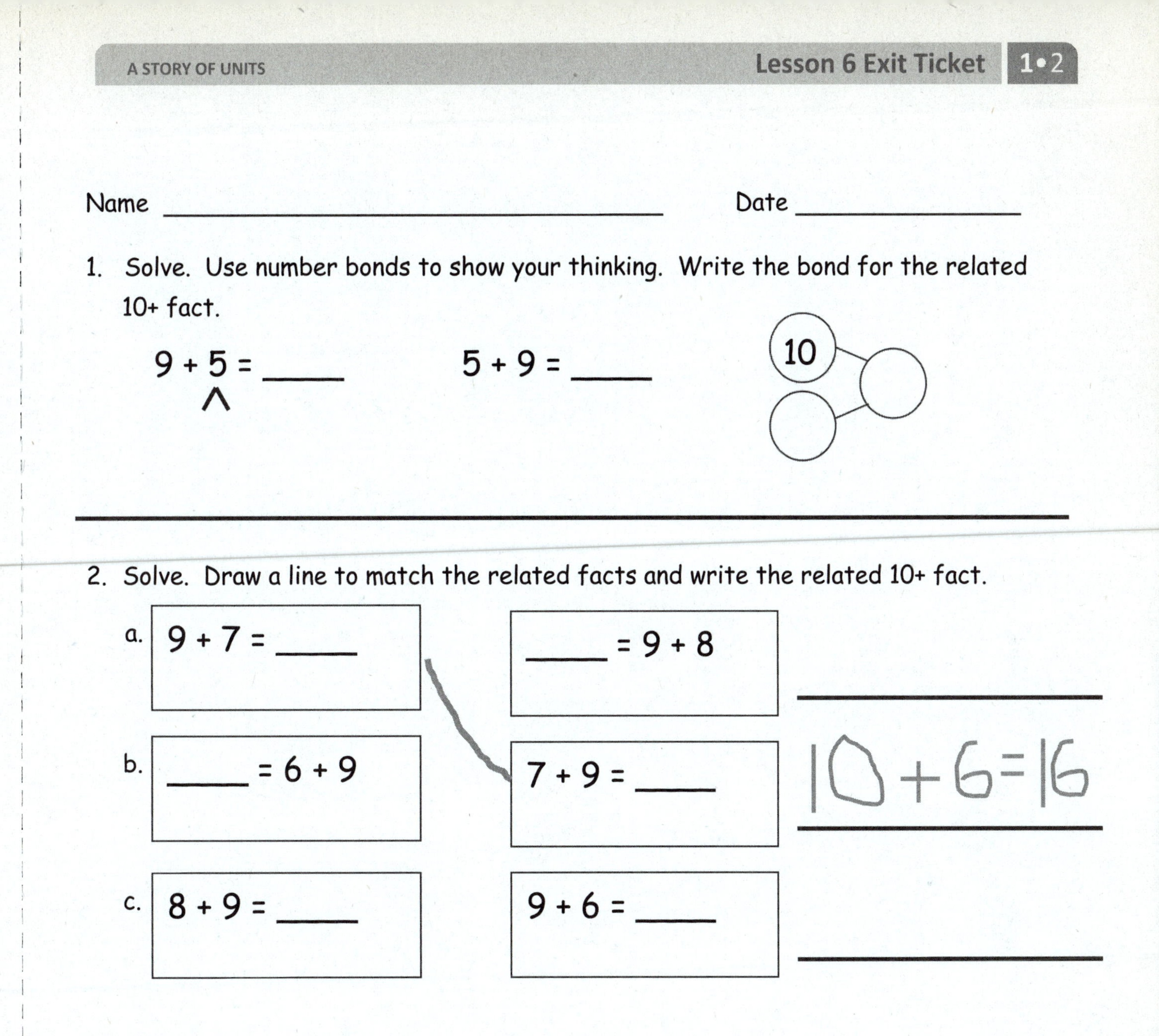

Name ______________________ Date ____________

1. Solve. Use number bonds to show your thinking. Write the bond for the related 10+ fact.

$9 + 5 =$ ____ $5 + 9 =$ ____

2. Solve. Draw a line to match the related facts and write the related 10+ fact.

a.	$9 + 7 =$ ____	____ $= 9 + 8$	____________
b.	____ $= 6 + 9$	$7 + 9 =$ ____	10 + 6 = 16
c.	$8 + 9 =$ ____	$9 + 6 =$ ____	____________

Read

Stacy made 6 drawings. Matthew made 2 drawings. Tim made 4 drawings. How many drawings did they make altogether? Use a drawing, a number sentence, and a statement to match the story.

Draw

Write

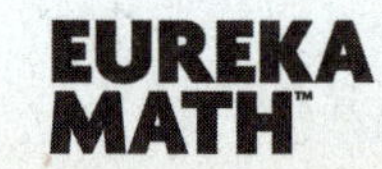

Name ______________________ Date __________

Circle to show how you made ten to help you solve.

1. John has 8 tennis balls. Toni has 5. How many tennis balls do they have in all?

John Toni

8 and ______ make ______.

10 and ______ make ______.

John and Toni have ______ tennis balls in all.

2. Bob has 8 raisins, and Jenny has 4. How many raisins do they have altogether?

8 and ______ make ______.

10 and ______ make ______.

Bob and Jenny have ______ raisins altogether.

3. There are 3 chairs on the right side of the classroom and 8 on the left side. How many total chairs are in the classroom?

8 and ______ make ______.

10 and ______ make ______.

There are _______ total chairs.

4. There are 7 children sitting on the rug and 8 children standing. How many children are there in all?

8 and ______ make ______.

10 and ______ make ______.

There are ______ children in all.

8 and ______ make ______.

10 and ______ make ______.

There are ______ children in all.

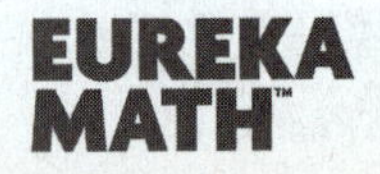

Name ______________________________ Date ______________

Draw, label, and circle to show how you made ten to help you solve.

Write the number sentences you used to solve.

Nick picks some peppers. He picks 5 green peppers and 8 red peppers. How many peppers does he pick in all?

8 and ______ make ______.

10 and ______ make ______.

Nick picks ____ peppers.

Read

A tree lost 8 leaves one day and 4 leaves the next. How many leaves did the tree lose at the end of the two days? Use a number bond, a number sentence, and a statement to match the story.

Extension: On the third day, the tree lost 6 leaves. How many leaves did it lose by the end of the third day?

Draw

Write

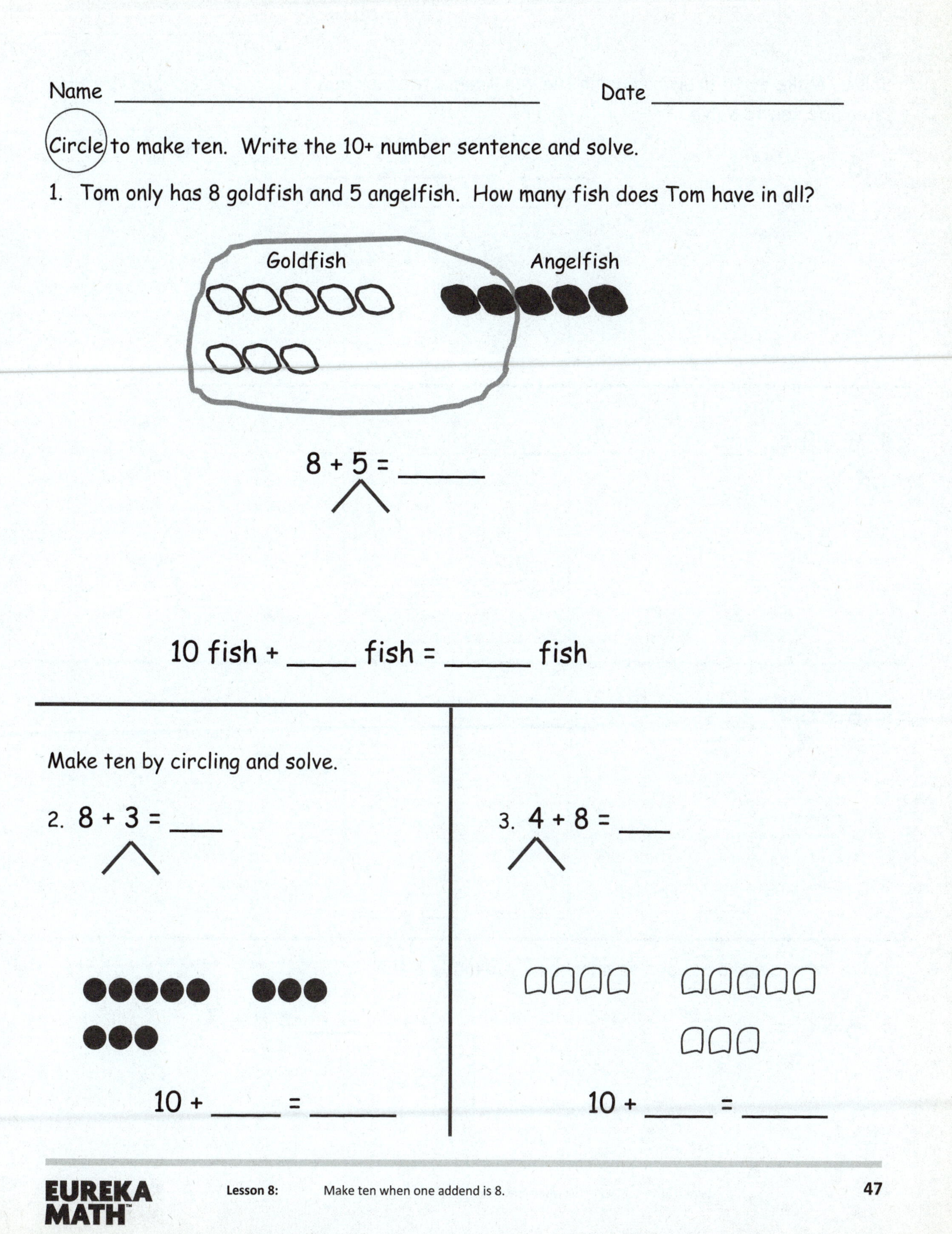

Name ______________________________ Date ______________

Circle to make ten. Write the 10+ number sentence and solve.

1. Tom only has 8 goldfish and 5 angelfish. How many fish does Tom have in all?

8 + 5 = _____

10 fish + _____ fish = _____ fish

Make ten by circling and solve.

2. 8 + 3 = ___

10 + _____ = _____

3. 4 + 8 = ___

10 + _____ = _____

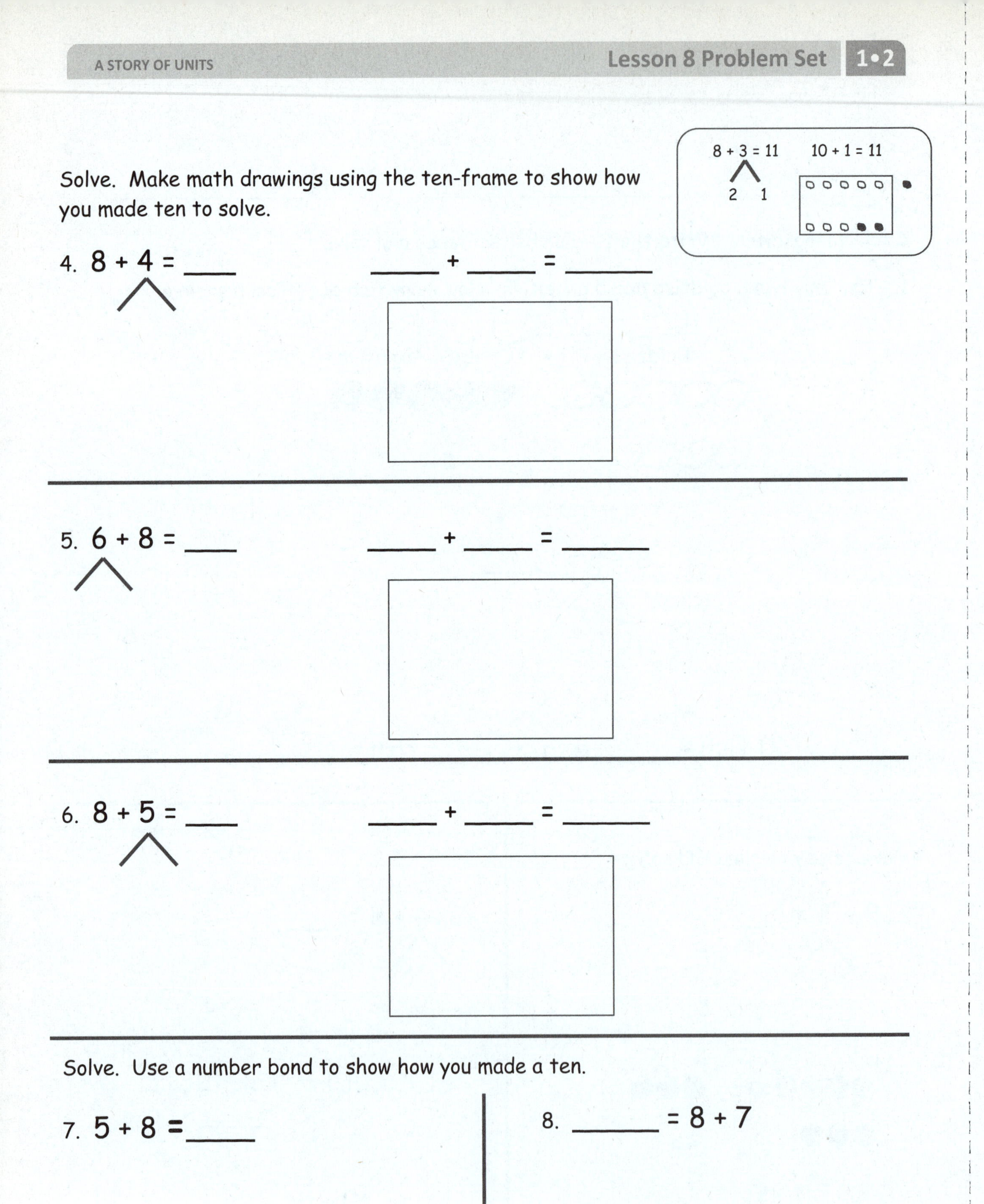

Solve. Make math drawings using the ten-frame to show how you made ten to solve.

4. 8 + 4 = ____ ____ + ____ = ____

5. 6 + 8 = ____ ____ + ____ = ____

6. 8 + 5 = ____ ____ + ____ = ____

Solve. Use a number bond to show how you made a ten.

7. 5 + 8 = ____

8. ____ = 8 + 7

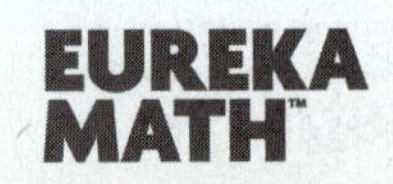

Name ______________________________ Date ______________

Make math drawings using the ten-frame to solve. Rewrite as a 10+ number sentence.

1. 6 + 8 = ____

10 + ____ = ____

2. ____ = 4 + 8

____ + ____ = ____

Read

A squirrel found 8 nuts in the morning, 5 nuts in the afternoon, and 2 nuts in the evening. How many nuts did the squirrel find in all?

Extension: The next day, the squirrel found 3 more nuts in the morning, 1 more in the afternoon, and 1 more in the evening. How many did he collect over the two days?

Draw

Write

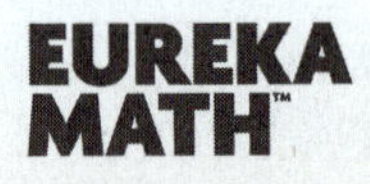

Name ______________________________ Date ______________

Make ten to solve. Use a number bond to show how you took 2 out to make ten.

1. Ben has 8 green grapes and 3 purple grapes. How many grapes does he have?

8 + 3 = _____   10 + _____ = _____

Ben has ___ grapes.

2. 8 + 4 = _____   10 + _____ = _____

Use number bonds to show your thinking. Write the 10+ fact.

3. 8 + 5 = _____   _____ + _____ = _____

4. 8 + 7 = _____   _____ + _____ = _____

5. 4 + 8 = _____   _____ + _____ = _____

6. 7 + 8 = _____   _____ + _____ = _____

7. 8 + _____ = 17   _____ + _____ = _____

Complete the addition sentences and number bonds.

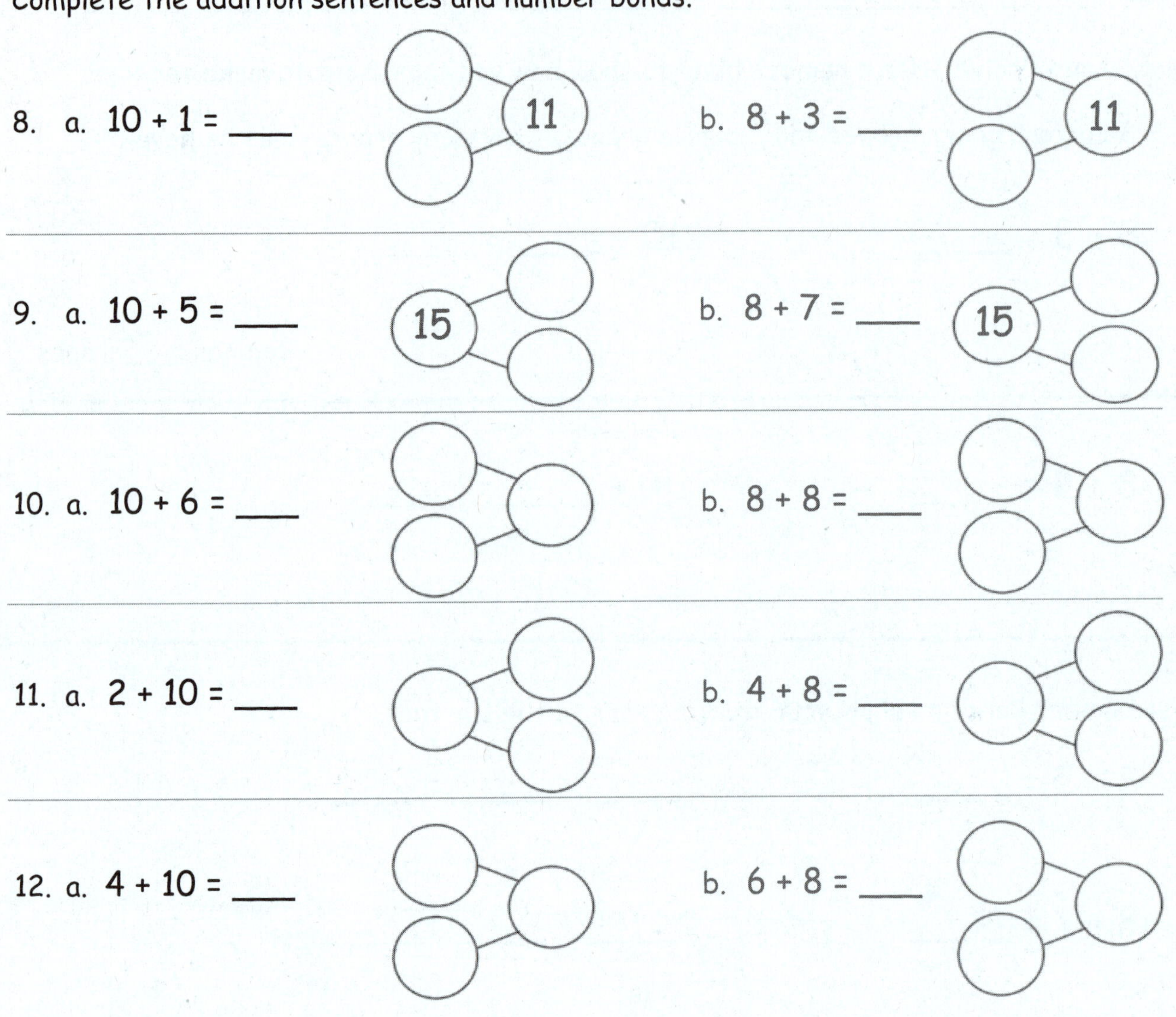

Name ______________________________ Date ______________

1. Seyla has 3 stamps in her collection. Her father gives her 8 more stamps. How many stamps does she have now? Show how you make ten, and write the 10+ fact.

3 + 8 = _____ 10 + _____ = _____

2. Complete the addition sentences and the number bonds.

a. 8 + 6 = ____

b. 10 + ____ = 14

Read

There were 4 boots by the classroom door, 8 boots in the hallway, and 6 boots by the teacher's desk. How many boots were there altogether?

Extension: How many pairs of boots were there in all?

Draw

Write

Name ______________________ Date ____________

Solve. Use number bonds or 5-group drawings if needed. Write the equal ten-plus number sentence.

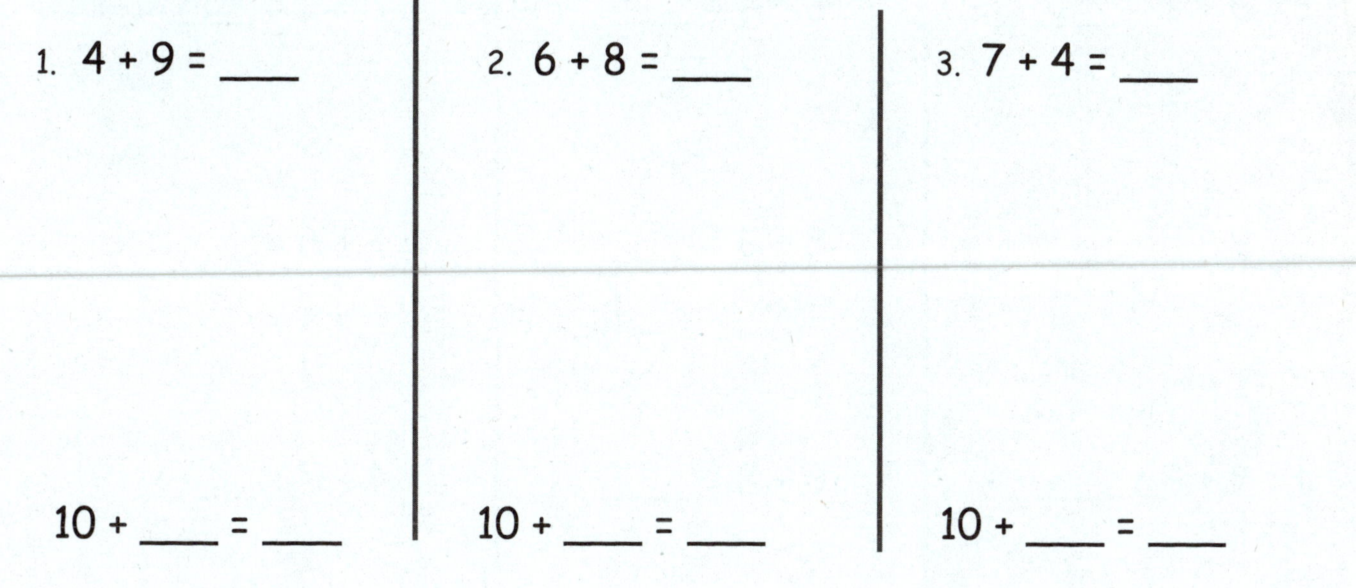

4. Match the equal expressions.

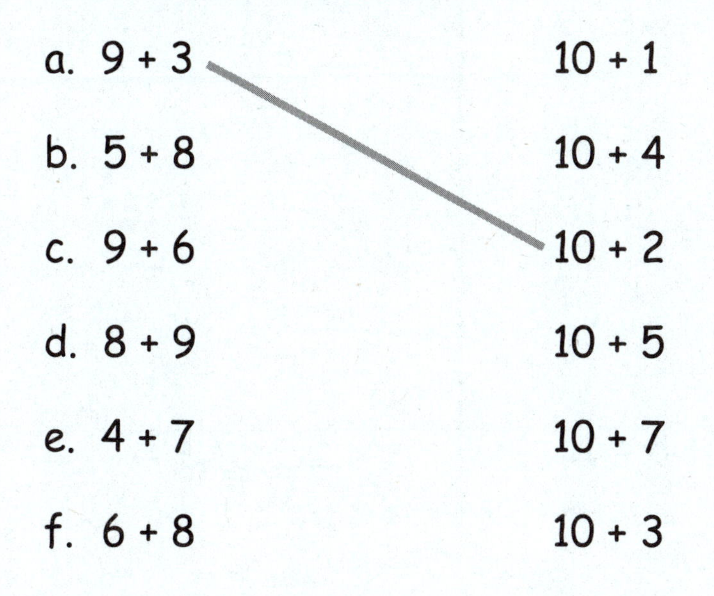

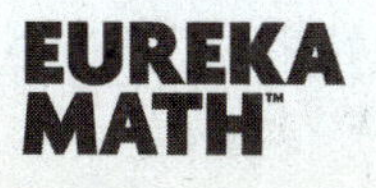

Complete the addition sentences to make them true.

	a.	b.	c.
5.	9 + 2 = ___	8 + 4 = ___	7 + 5 = ___
6.	9 + 5 = ___	8 + 3 = ___	7 + 6 = ___
7.	6 + 9 = ___	6 + 8 = ___	4 + 7 = ___
8.	7 + 9 = ___	5 + 8 = ___	7 + 7 = ___
9.	9 + ___ = 17	8 + ___ = 16	7 + ___ = 16
10.	___ + 9 = 15	___ + 8 = 15	___ + 7 = 17

Name ______________________________ Date ______________

Solve. Use number bonds or 5-group drawings if needed. Write the equal ten-plus number sentence.

a.	b.	c.
9 + 5 = ___	8 + 4 = ___	7 + 6 = ___
10 + ___ = ___	10 + ___ = ___	10 + ___ = ___

Read

Nicholas bought 9 green apples and 7 red apples. Sofia bought 10 red apples and 6 green apples. Sofia thinks she has more apples than Nicholas. Is she right? Choose a strategy you have learned to show your work. Then, write number sentences to show how many apples Nicholas and Sofia each have.

Draw

Write

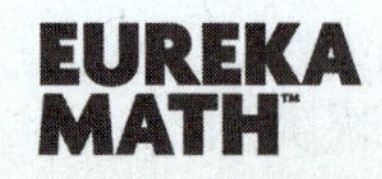

Name ______________________________ Date ______________

Jeremy had 7 big rocks and 8 little rocks in his pocket.

How many rocks does Jeremy have?

1. Circle all student work that correctly matches the story.

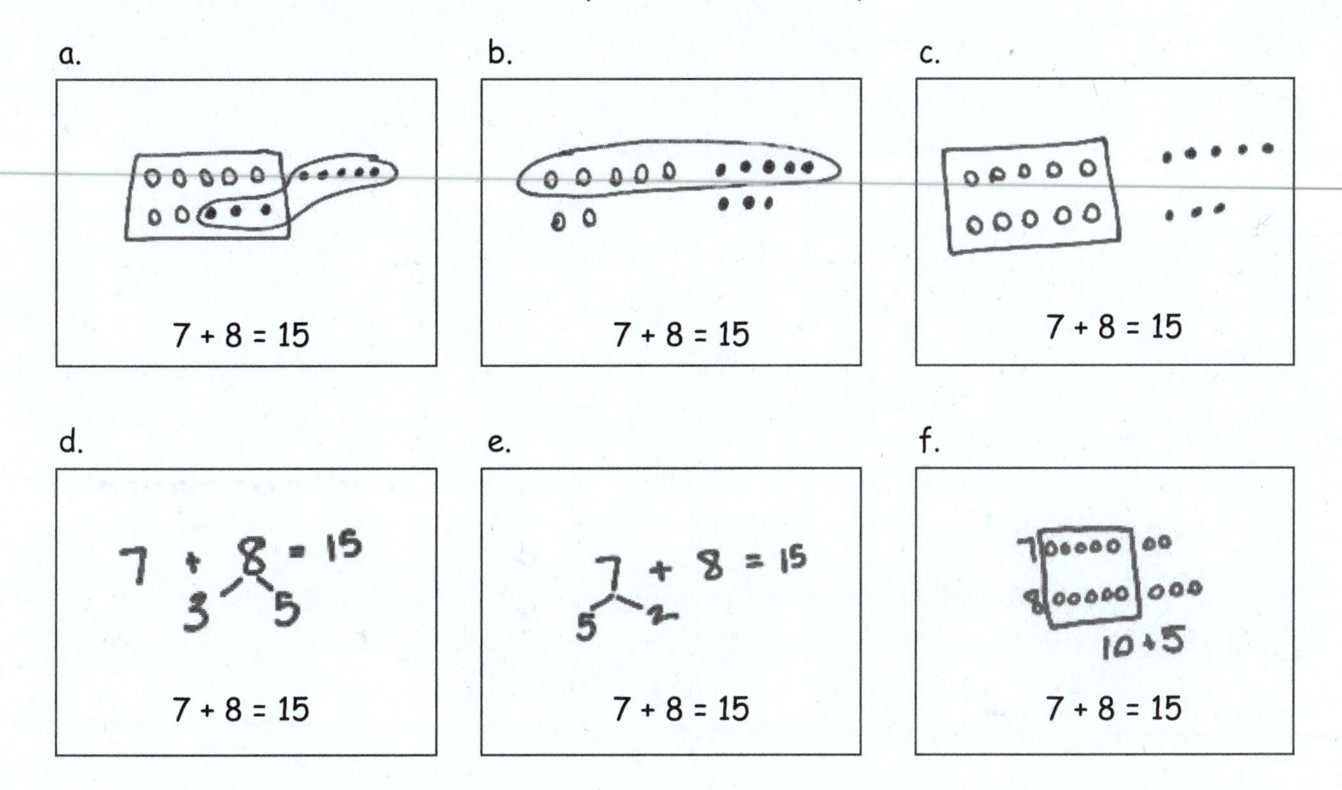

2. Fix the work that was incorrect by making a new drawing in the space below with the matching number sentence.

Solve on your own. Show your thinking by drawing or writing. Write a statement to answer the question.

3. There are 4 vanilla cupcakes and 8 chocolate cupcakes for the party. How many cupcakes were made for the party?

4. There are 5 girls and 7 boys on the playground. How many students are on the playground?

When you are done, share your solutions with a partner. How did your partner solve each problem? Be ready to share how your partner solved the problems.

Name ______________________________ Date ______________

John thinks the problem below should be solved using 5-group drawings, and Sue thinks it should be solved using a number bond. Solve both ways, and circle the strategy you think is the more efficient.

Kim scores 5 goals in her soccer game and 8 runs in her softball game. How many points does she score altogether?

John's Work

Sue's Work

Read

Claudia bought 8 red apples and 9 green apples. How many apples does Claudia have altogether? Make a math drawing, number sentence, and statement to show your thinking.

Extension: Claudia ate 3 red apples, and her friend ate 4 green apples. How many apples does Claudia have now?

Draw

Write

Name ______________________________ Date ______________

Make a simple math drawing. Cross out from the 10 ones or the other part in order to show what happens in the stories.

1. Bill has 16 grapes. 10 are on one vine, and 6 are on the ground.
   Bill eats 9 grapes from the vine. How many grapes does Bill have left?

16
10
6

Bill has ____ grapes now.

---

2. 12 frogs are in the pond. 10 are on a lily pad, and 2 are in the water. 9 frogs hop off the lily pad and out of the pond. How many frogs are in the pond?

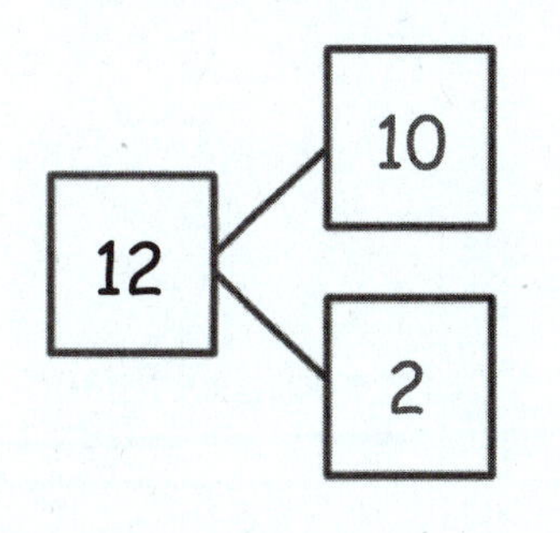

There are ____ frogs still in the pond.

---

3. Kim has 14 stickers. 10 stickers are on the first page, and 4 stickers are on the second page. Kim loses 9 stickers from the first page. How many stickers are still in her book?

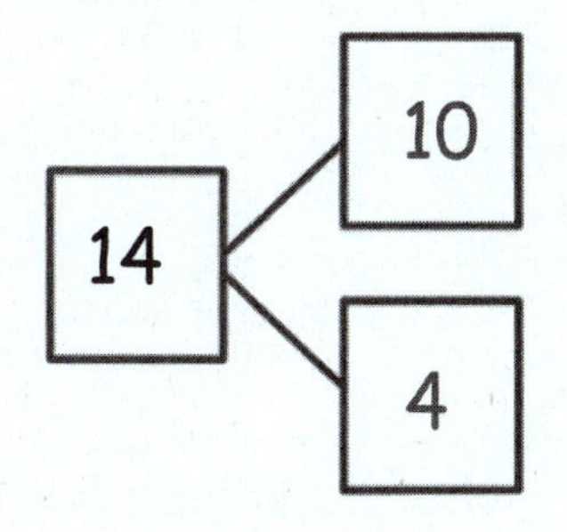

Kim has ____ stickers in her book.

4. 10 eggs are in a carton, and 5 eggs are in a bowl. Joe's father cooks 9 eggs from the carton. How many eggs are left?

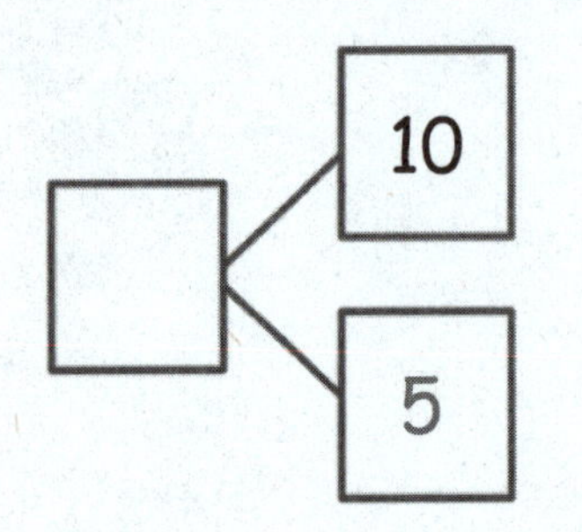

There are ___ eggs left.

---

5. Jana had 10 wrapped gifts on the table and 7 wrapped gifts on the floor. She unwrapped 9 gifts from the table. How many gifts are still wrapped?

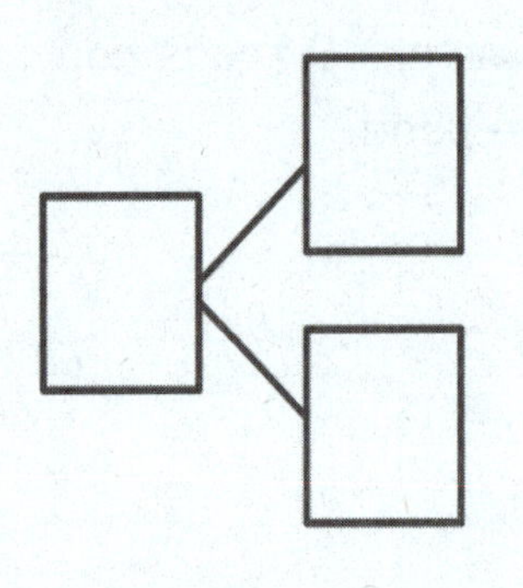

Jana has ___ gifts still wrapped.

---

6. There are 10 cupcakes on a tray and 8 on the table. On the tray, there are 9 vanilla cupcakes. The rest of the cupcakes are chocolate. How many cupcakes are chocolate?

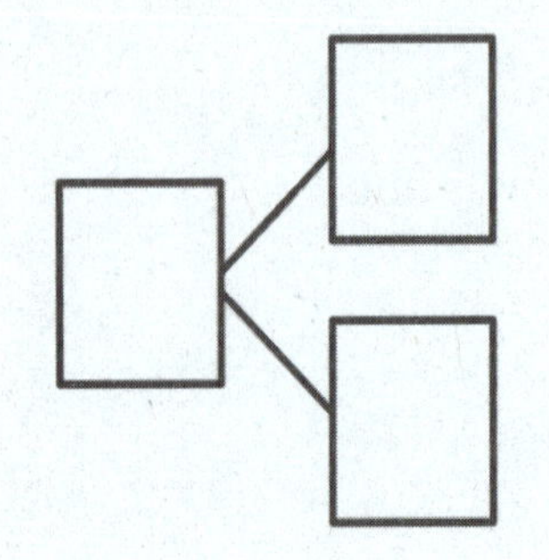

There are ___ chocolate cupcakes.

Name ______________________________ Date ________________

Make a simple math drawing. Cross out from the 10 ones to show what happens in the story.

There were 16 books on the table. 10 books were about dinosaurs. 6 books were about fish. A student took 9 of the dinosaur books. How many books were left on the table?

There were ____ books left on the table.

00000 00000

5-group row insert

Read

Ten snowflakes fell on Sam's mitten, and 6 fell on his coat. Nine of the snowflakes on Sam's mitten melted. How many snowflakes are left? Write a subtraction sentence to show how many snowflakes are left.

Draw

Write

Name ______________________________ Date ______________

Solve. Use 5-group rows, and cross out to show your work.

1. Mike has 10 cookies on a plate and 3 cookies in a box. He eats 9 cookies from the plate. How many cookies are left?

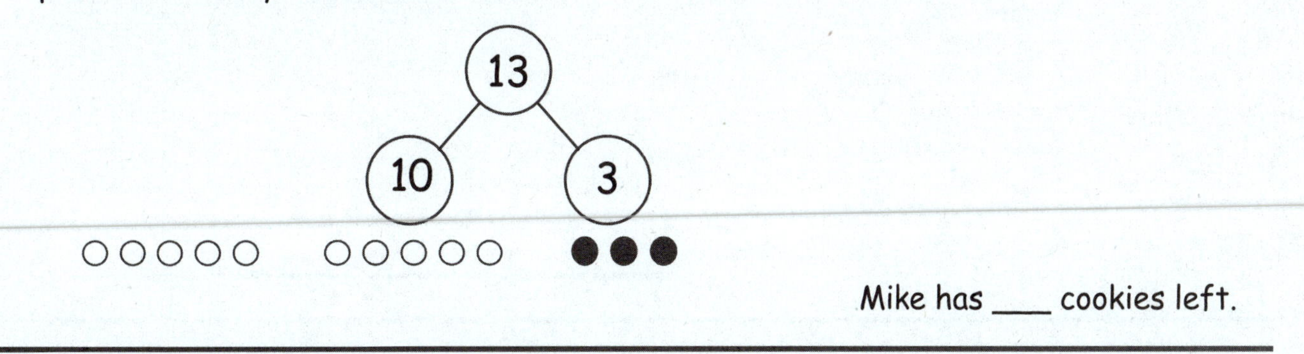

Mike has ___ cookies left.

2. Fran has 10 crayons in a box and 5 crayons on the desk. Fran lends Bob 9 crayons from the box. How many crayons does Fran have to use?

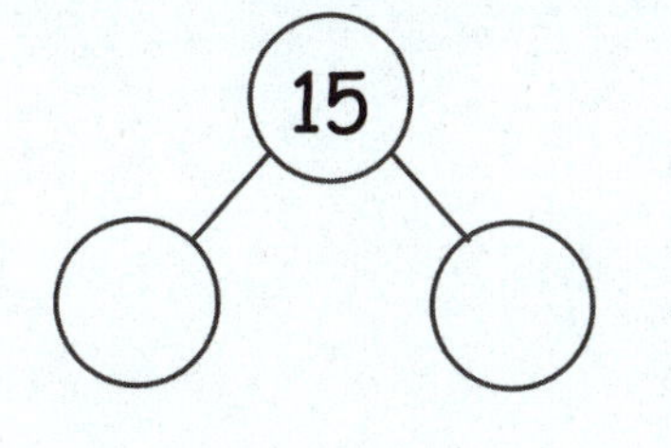

Fran has ___ crayons to use.

3. 10 ducks are in the pond, and 7 ducks are on the land. 9 of the ducks in the pond are babies, and all the rest of the ducks are adults. How many adult ducks are there?

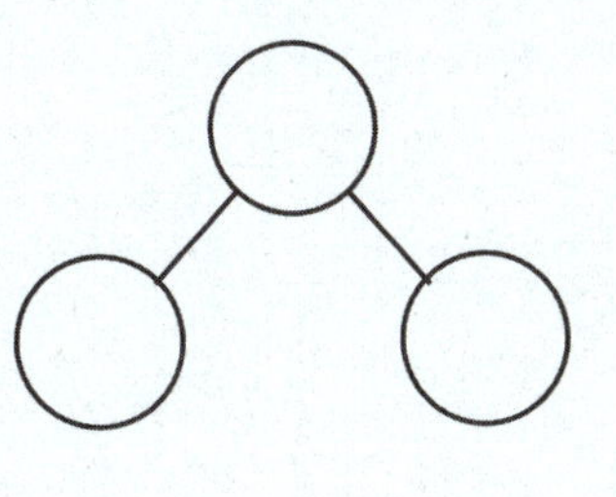

There are ___ adult ducks.

With a partner, create your own stories to match, and solve the number sentences. Make a number bond to show the whole as 10 and some ones. Draw 5-group rows to match your story. Write the complete number sentence on the line.

4. 16 - 9 = ☐

____

5. 12 - 9 = ☐

____

6. 19 - 9 = ☐

____

Name ______________________________ Date ______________

Solve. Fill in the number bond. Use 5-group rows, and cross out to show your work.

Gabriela has 4 hair clips in her hair and 10 hair clips in her bedroom. She gives 9 of the hair clips in her room to her sister. How many hair clips does Gabriela have now?

Gabriela has ___ hair clips.

Read

Sarah has 6 blue beads in her bag and 4 green beads in her pocket.

She gives away the 6 blue beads and 3 green beads. How many beads does she have left?

Draw

Write

Name ______________________ Date ____________

1. Match the pictures with the number sentences.

a. 11 - 9 = 2

b. 14 - 9 = 5

c. 16 - 9 = 7

d. 18 - 9 = 9

e. 17 - 9 = 8

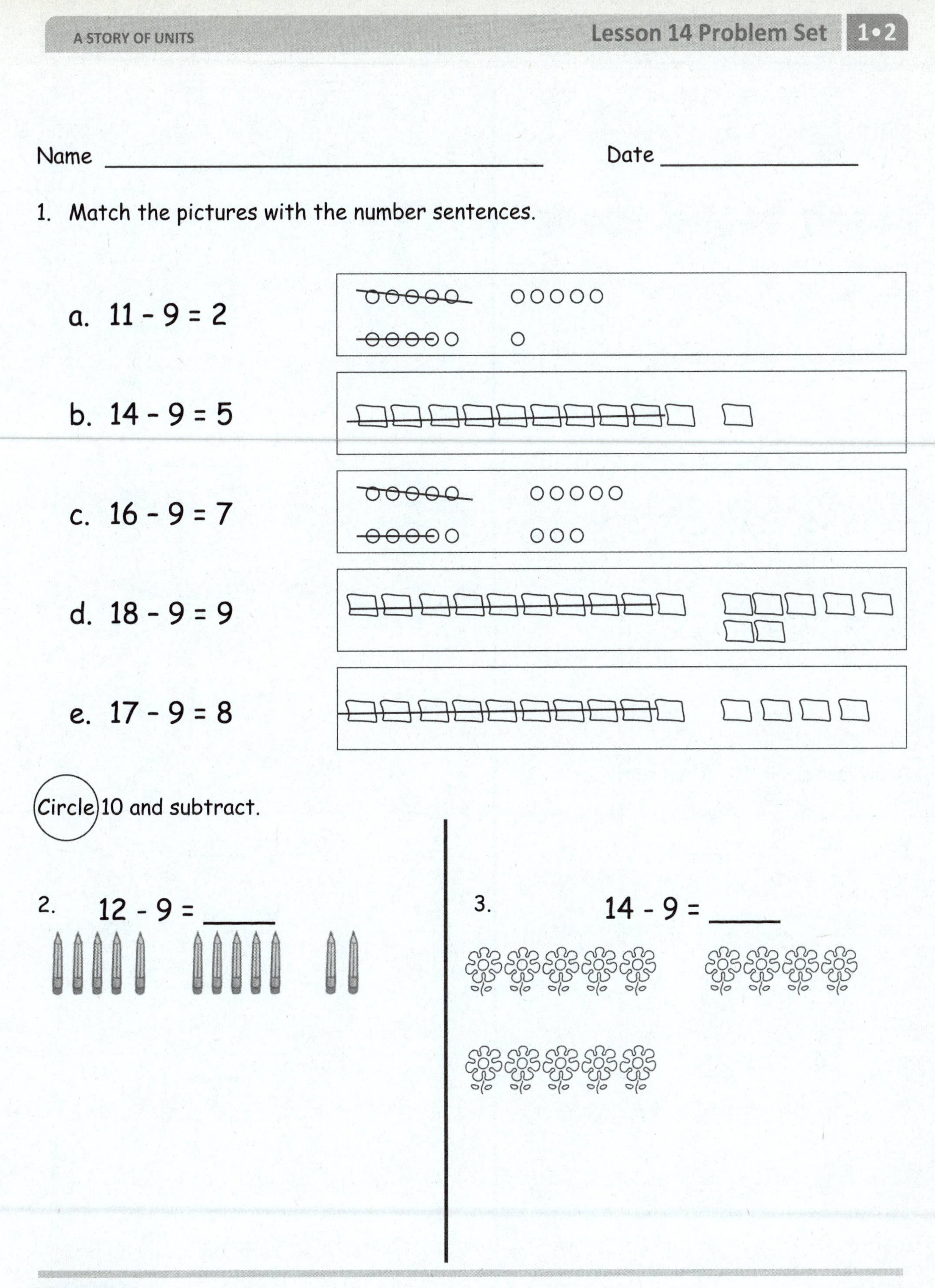

Circle 10 and subtract.

2. 12 - 9 = ____

3. 14 - 9 = ____

4. 15 - 9 = ____

5. 13 - 9 = ____

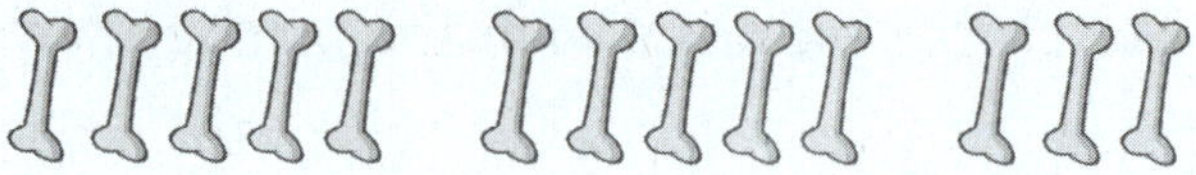

6. 16 - 9 = ____

7. 17 - 9 = ____

Draw and circle 10. Then subtract.

8. 12 - 9 = ____

9. 13 - 9 = ____

10. 14 - 9 = ____

11. 15 - 9 = ____

Name ______________________________ Date ______________

Draw and circle 10. Solve and make a number bond.

1. 17 - 9 = _____

2. 14 - 9 = _____

3. 15 - 9 = ____

4. 18 - 9 = ____

Read

Julian has 7 markers. His mother gives him 8 more. He loses 9 markers.

How many does he have left?

Draw

Write

Name ______________________________ Date ______________

1. Match the pictures with the number sentences.

a. 13 - 9 = 4

b. 14 - 9 = 5

c. 17 - 9 = 8

d. 18 - 9 = 9

e. 16 - 9 = 7

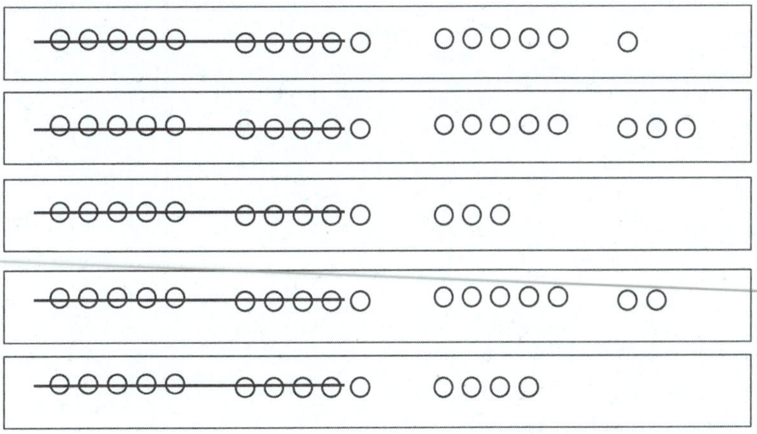

Draw 5-group rows. Visualize and then cross out to solve. Complete the number sentences.

2. 11 - 9 = _____

3. 13 - 9 = _____

4. 16 - 9 = _____

5. 17 - 9 = _____

6. $14 - 9 =$ _____

7. $13 - 9 =$ _____

8. $12 - 9 =$ _____

9. $15 - 9 =$ _____

10. Show making 10 and taking from 10 to complete the two number sentences.

a. $5 + 9 =$ ____

b. $14 - 9 =$ ____

11. Make a number bond for Problem 10. Write two additional number sentences that use this number bond.

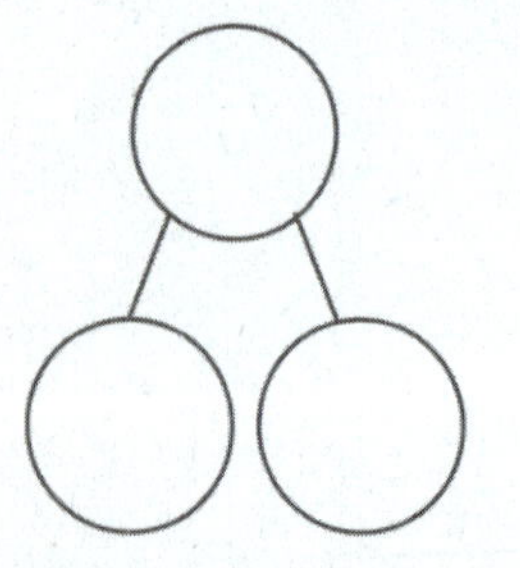

______________________ ______________________

Name ______________________________ Date ______________

Draw 5-group rows, and cross out to solve. Complete the number sentences.

1. 17 - 9 = ____

2. 19 - 9 = ____

Read

There were 16 coats on the rack. Nine students took their coats to go outside. How many coats were still on the rack?

Extension: If 4 more students take their coats to go outside, how many coats will still be hanging?

Draw

Write

Name ______________________________ Date ______________

Solve the problem by counting on (a) and using a number bond to take from ten (b).

1. Lucy had 12 balloons at her birthday party. She gave 9 balloons to her friends. How many balloons did she have left?

   a. $12 - 9 =$ _____

   b. $12 - 9 =$ _____

   Lucy had ___ balloons left.

---

2. Justin had 15 blueberries on his plate. He ate 9 of them. How many does he have left to eat?

   a. $15 - 9 =$ _____

   b. $15 - 9 =$ _____

   Justin has ___ blueberries left to eat.

Complete the subtraction sentences by using the take from ten strategy and counting on. Tell which strategy you would prefer to use for Problems 3 and 4.

3. a. 11 - 9 = ___ b. 11 - 9 = ___

☐ take from ten

☐ count on

---

4. a. 18 - 9 = ___ b. 18 - 9 = ___

☐ take from ten

☐ count on

---

5. Think about how to solve the following subtraction problems:

16 - 9	12 - 9	18 - 9
11 - 9	15 - 9	14 - 9
13 - 9	19 - 9	17 - 9

Choose which problems you think are easier to count on from 9 and which are easier to use the take from ten strategy. Write the problems in the boxes below.

Problems to use the *count on* strategy with:	**Problems to use the *take from ten* strategy with:**

Were there any problems that were just as easy using either method? Did you use a different method for any problems?

Name ______________________________ Date ______________

Complete the subtraction sentences by using both the count on and take from ten strategies.

1. a. 13 - 9 = ___ b. 13 - 9 = ___

2. a. 17 - 9 = ___ b. 17 - 9 = ___

Read

Gisella had 13 markers in her bag. Eight markers fell out of the bag.

How many markers does Gisella have now?

Draw

Write

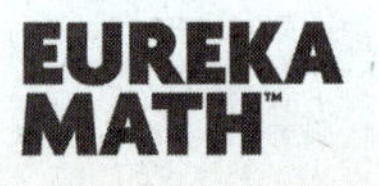

Name ______________________________ Date ______________

1. Match the pictures with the number sentences.

a. 12 - 8 = 4

b. 17 - 8 = 9

c. 16 - 8 = 8

d. 18 - 8 = 10

e. 14 - 8 = 6

Circle 10 and subtract.

2. 13 - 8 = _____

3. 11 - 8 = _____

4. 15 - 8 = _____

5. 19 - 8 = _____

6. 16 - 8 = _____

7. 17 - 8 = _____

Draw and circle 10, **or** break apart the teen number with a number bond. Then subtract.

8. 12 - 8 = ____

9. 13 - 8 = ____

10. 14 - 8 = ____

11. 15 - 8 = ____

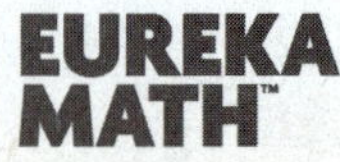

Name ________________________ Date ____________

1. Draw and circle 10. Then subtract.

a. 12 - 8 = _____	b. 14 - 8 = _____

2. Use a number bond to break apart the teen number. Then subtract.

15 - 8 = _____

Read

Juliana rolls 8 cars down a ramp. If she started with 15 cars at the top of the ramp, how many cars does Juliana still have at the top of the ramp?

Draw

Write

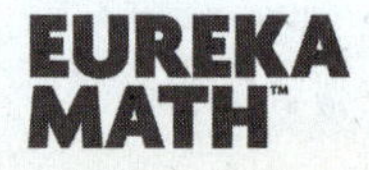

Name ______________________________ Date ______________

1. Match the pictures with the number sentences.

a. 13 - 8 = 5

b. 14 - 8 = 6

c. 17 - 8 = 9

d. 18 - 8 = 10

e. 16 - 8 = 8

Make a math drawing of a 5-group row and some ones to solve the following problems. Write the addition sentence that shows how to add the parts after subtracting 8 or 9.

2. 11 - 8 = _____ ______________________

3. 12 - 8 = _____ ______________________

4. 15 - 8 = _____ ______________________

5. 19 - 8 = _____ ______________________

6. 16 - 8 = _____ ______________________

7. 16 - 9 = _____ ______________________

8. 14 - 9 = _____ ______________________

9. Show how to make ten and take from ten to solve the two number sentences.

a. 6 + 8 = _____

b. 14 - 8 = _____

Name ______________________________ Date ______________

Draw 5-group rows, and cross out to solve. Complete the number sentences. Write the 2+ addition sentence that helped you add the two parts.

1. 14 - 8 = ____

2 + ____ = ____

---

2. 17 - 8 = ____

2 + ____ = ____

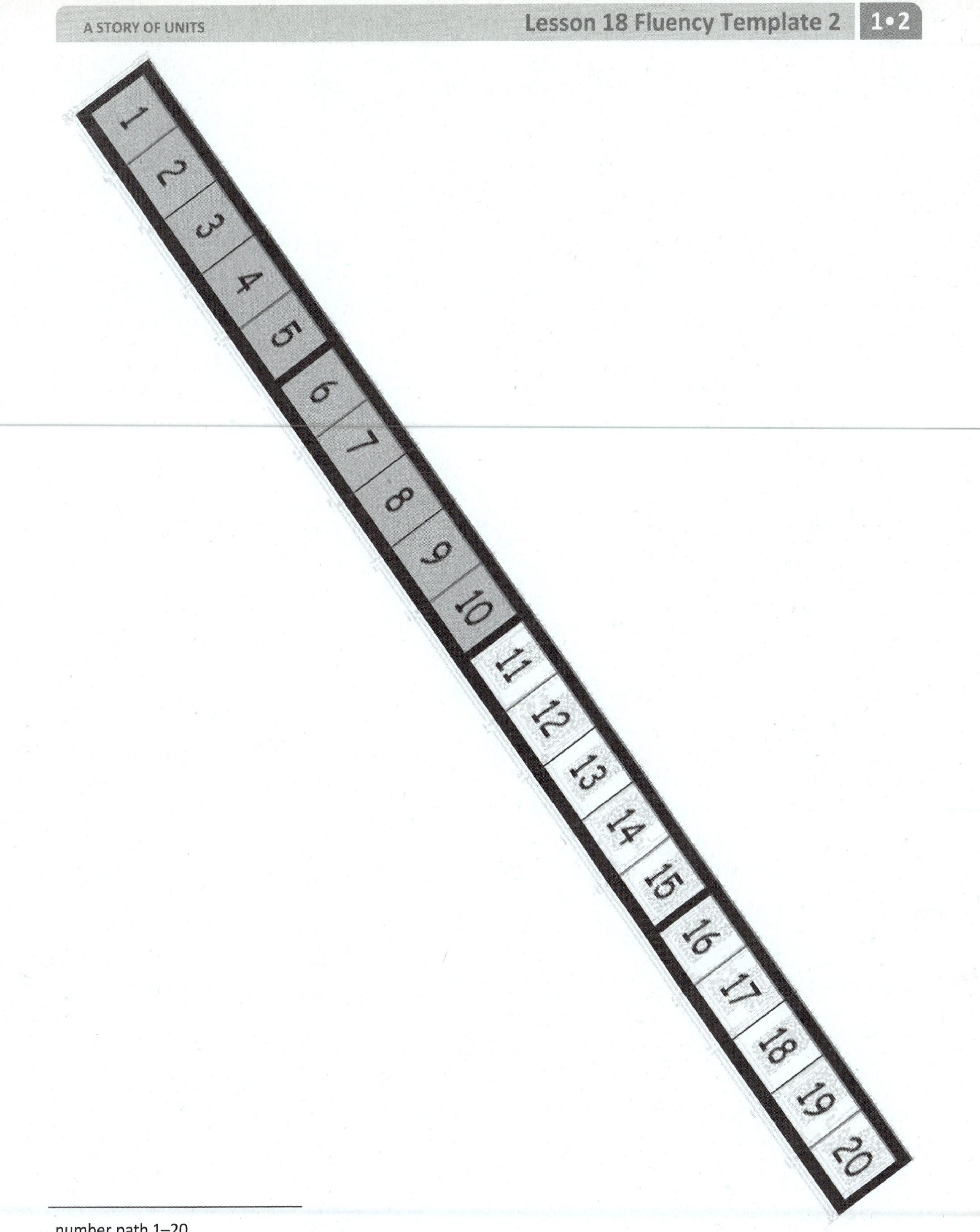

number path 1–20

Read

Carla, Jose, and Yannis each have 8 cherries.

They all get more cherries to put in their bowls.

Now, Carla has 12 cherries, Jose has 14 cherries, and Yannis has 16 cherries.

How many more cherries did they each put in their bowls?

Write a number sentence for each answer.

Draw

Write

Name ______________________________ Date ______________

Use a number bond to show how you used the take from ten strategy to solve the problem.

1. Kevin had 14 crayons. Eight of the crayons were broken. How many of his crayons were not broken?

$14 - 8 =$ _____

14 - 8
10 4
Subtract 8 from 10.
2 and 4 is 6.

Kevin had ___ crayons that were not broken.

Use number bonds to show your thinking.

2. $17 - 8 =$ _____

3. $18 - 8 =$ _____

Count on to solve.

4. $13 - 8 =$ _____

5. $15 - 8 =$ _____

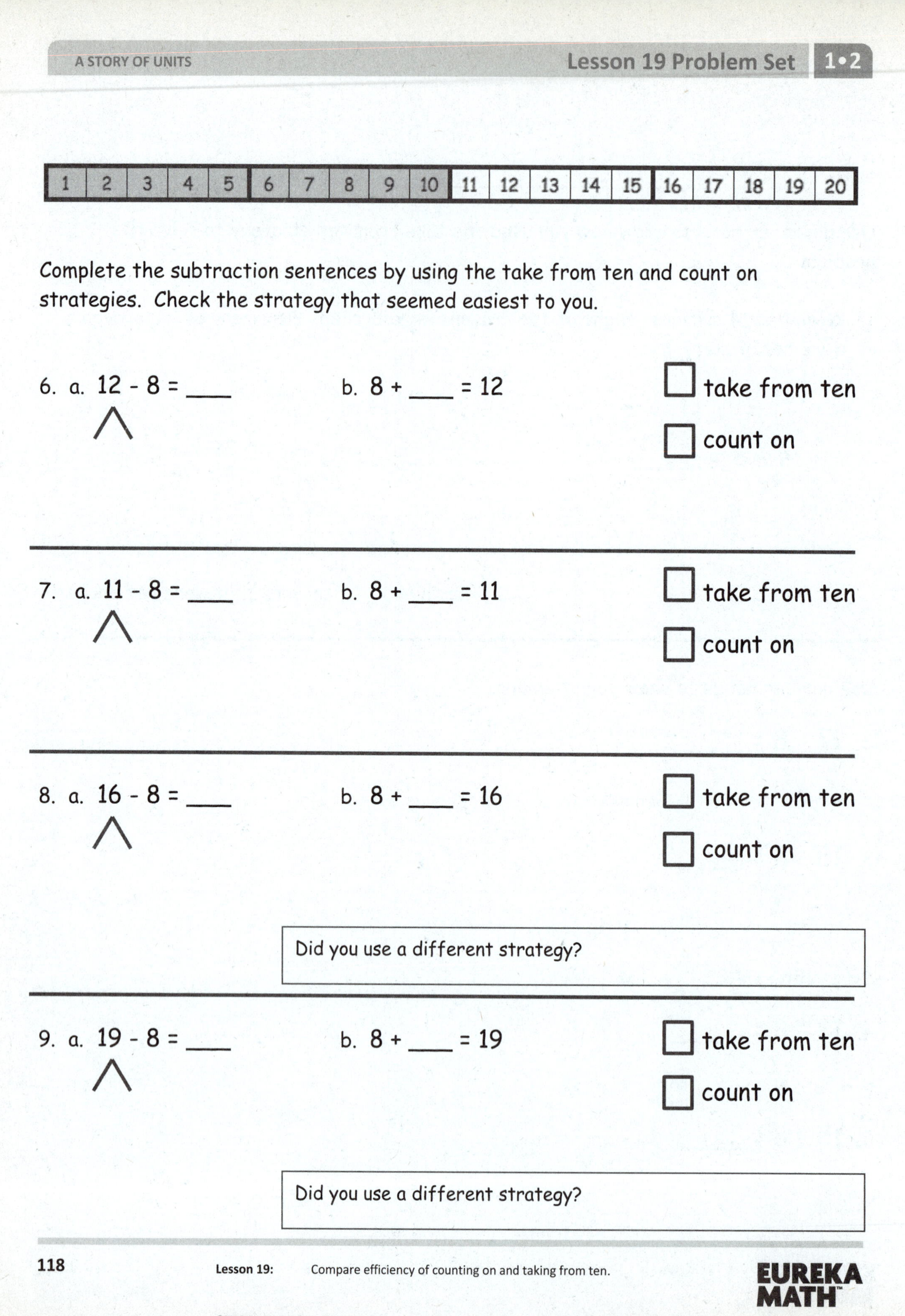

Complete the subtraction sentences by using the take from ten and count on strategies. Check the strategy that seemed easiest to you.

6. a. 12 - 8 = ____   b. 8 + ____ = 12

☐ take from ten
☐ count on

7. a. 11 - 8 = ____   b. 8 + ____ = 11

☐ take from ten
☐ count on

8. a. 16 - 8 = ____   b. 8 + ____ = 16

☐ take from ten
☐ count on

Did you use a different strategy?

9. a. 19 - 8 = ____   b. 8 + ____ = 19

☐ take from ten
☐ count on

Did you use a different strategy?

Name ______________________________ Date ______________

Complete the subtraction sentences by using the take from ten strategy and count on.

1	2	3	4	5	6	7	8	9	10	11	12	13	14	15	16	17	18	19	20

1. a. 11 - 8 = ____ b. 8 + ____ = 11

2. a. 15 - 8 = ____ b. 8 + ____ = 15

Read

Imran has 8 crayons in his pencil box and 7 crayons in his desk.

How many crayons does Imran have in total?

Draw

Write

Name ________________________ Date ____________

Solve the problems below. Use drawings or number bonds.

1. 11 - 9 = _____

2. 11 - 8 = _____

3. 13 - 9 = _____

4. 13 - 8 = _____

5. 13 - 7 = _____

6. 12 - 7 = _____

7. Match the equal expressions.

a. 16 - 7 13 - 9
b. 17 - 7 18 - 9
c. 12 - 8 15 - 9
d. 14 - 8 18 - 8

Complete the subtraction sentences to make them true.

	a.	b.	c.
8.	12 - 9 = ___	13 - 9 = ___	14 - 9 = ___
9.	12 - 8 = ___	13 - 8 = ___	14 - 8 = ___
10.	11 - 7 = ___	12 - 7 = ___	13 - 7 = ___
11.	16 - 9 = ___	18 - 9 = ___	17 - 9 = ___
12.	16 - ___ = 9	15 - ___ = 9	15 - ___ = 7
13.	15 - ___ = 6	11 - ___ = 3	16 - ___ = 7

Name ______________________________ Date ______________

Solve the problems below. Use drawings or number bonds.

a. 14 - 9 = ____ b. 14 - 7 = ____ c. 14 - 8 = ____

---

d. 16 - 7 = ____ e. 16 - 9 = ____ f. 16 - 8 = ____

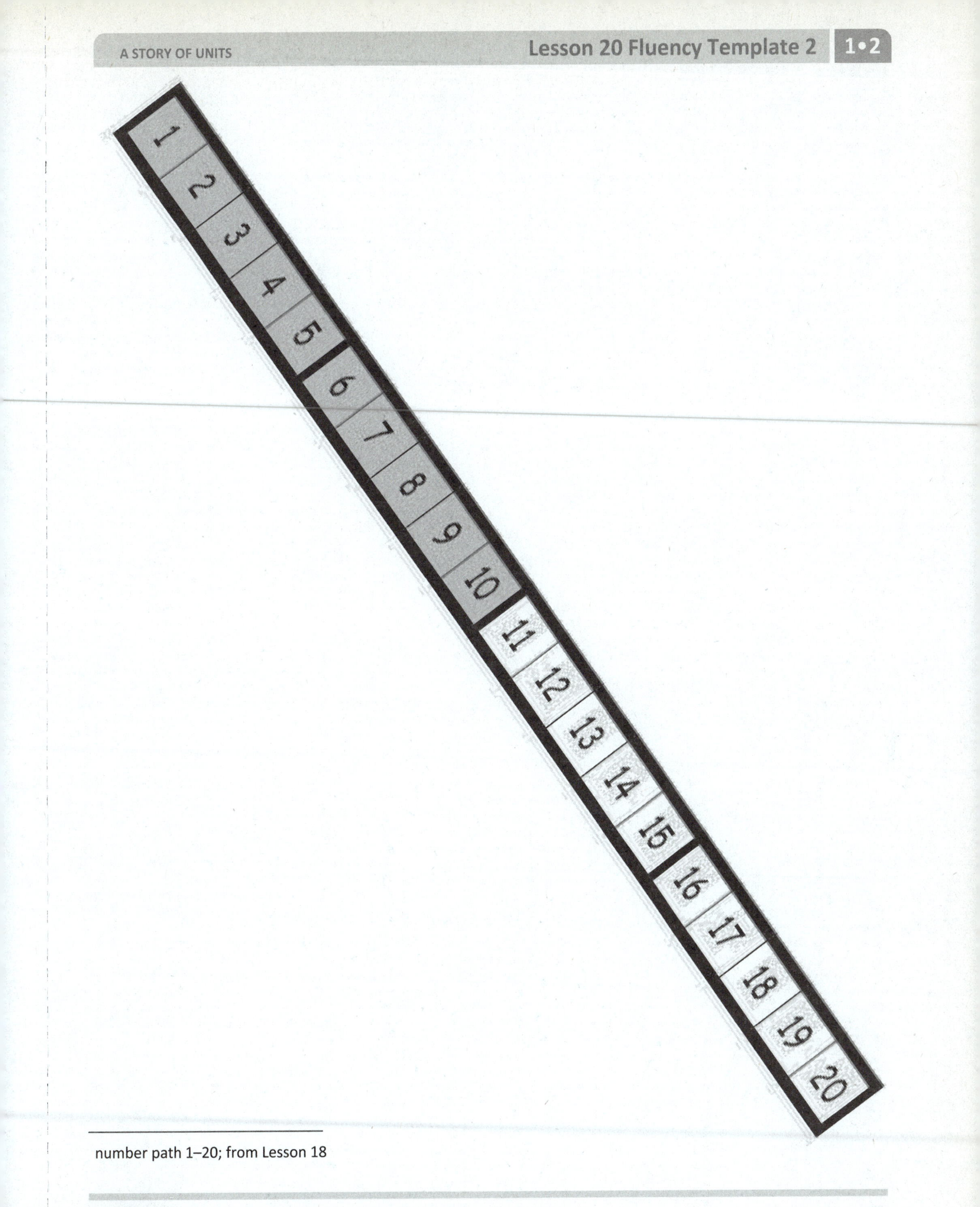

number path 1–20; from Lesson 18

Read

There are 16 reading mats in the classroom. If 9 reading mats are being used, how many reading mats are still available?

Draw

Write

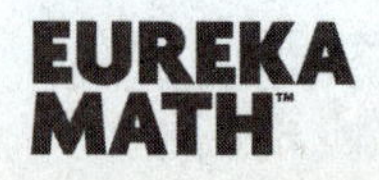

Name ____________________ Date ____________

There were 16 dogs playing at the park. Seven of the dogs went home.
How many of the dogs are still at the park?

1. Circle all the student work that correctly matches the story.

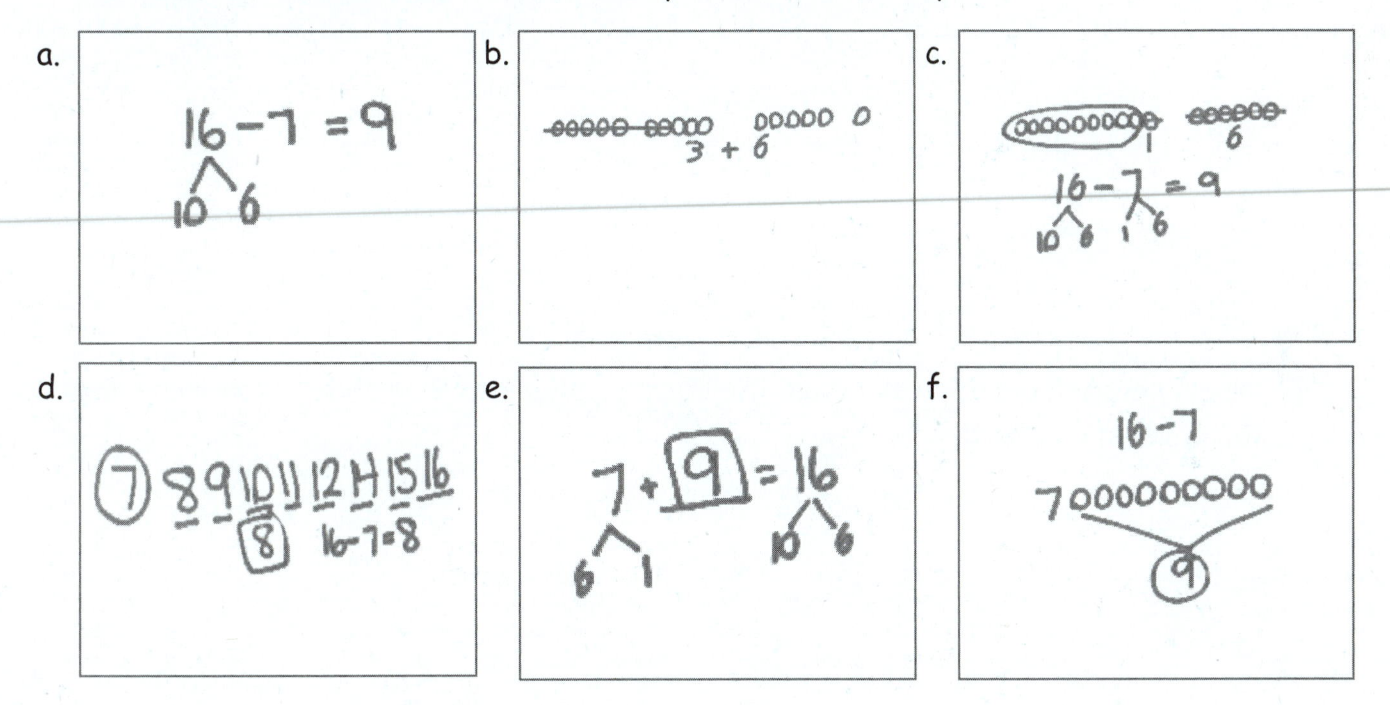

2. Fix the work that was incorrect by making a new drawing in the space below with the matching number sentence.

Solve on your own. Show your thinking by drawing or writing.
Write a statement to answer the question.

3. There were 12 sugar cookies in the box. My friend and I ate 5 of them. How many cookies are left in the box?

4. Megan checked out 17 books from the library. She read 9 of them. How many does she have left to read?

When you are done, share your solutions with a partner. How did your partner solve each problem? Be ready to share how your partner solved the problem.

Name ______________________________ Date ______________

Meg thinks using the take from ten strategy is the best way to solve the following word problem. Bill thinks that solving the problem using the count on strategy is a better way. Solve both ways, and explain which strategy you think is best.

Strategies:

- Take from 10
- Make 10
- Count on
- I just knew

Mike and Sally have 6 cats. They have 14 pets in all. How many pets do they have that are *not* cats?

Meg's strategy	Bill's strategy

I think ______________ strategy is best because ______________

____

____.

Name ______________________________ Date ______________

Read the word problem.
Draw and label.
Write a number sentence and a statement that matches the story.

1. This week, Maria ate 5 yellow plums and some red plums. If she ate 11 plums in all, how many red plums did Maria eat?

2. Tatyana counted 14 frogs. She counted 8 swimming in the pond and the rest sitting on lily pads. How many frogs did she count sitting on lily pads?

3. Some children are on the playground. Eight are on the swings, and the rest are playing tag. There are 15 children in all. How many children are playing tag?

---

4. Oziah read some non-fiction books. Then, he read 7 fiction books. If he read 16 books altogether, how many non-fiction books did Oziah read?

Meet with a partner, and share your drawings and sentences.
Talk with your partner about how your drawing matches the story.

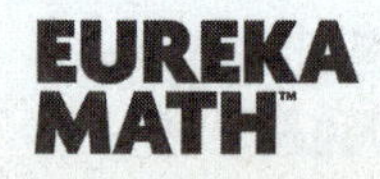

Name ______________________________ Date ______________

Read the word problem.
Draw and label.
Write a number sentence and a statement that matches the story.

Remember to draw a box around your solution in the number sentence.

1. Some students in Mrs. See's class are walkers. There are 17 students in her class in all. If 8 students ride the bus, how many students are walkers?

2. I baked 13 loaves of bread for a party. Some were burnt, so I threw them away. I brought the remaining 8 loaves to the party. How many loaves of bread were burnt?

Read

In the morning, there were 8 leaves on the floor under the ficus tree. During the day, more leaves fell on the floor. Now, there are 13 leaves on the floor. How many leaves fell during the day?

Draw

Write

Name ______________________________ Date ______________

Read the word problem.
Draw and label.
Write a number sentence and a statement that matches the story.

1. Janet read 8 books during the week. She read some more books on the weekend. She read 12 books total. How many books did Janet read on the weekend?

---

2. Eric s[illegible]d 13 goals this season! He scored 5 goals before the playoffs. How many goals [illegible] Eric score during the playoffs?

3. There were 8 ladybugs on a branch. Some more came. Then, there were 15 ladybugs on the branch. How many ladybugs came?

---

4. Marco's friend gave him some baseball cards at school. If he was already given 9 baseball cards by his family, and he now has 19 cards in all, how many baseball cards did he get in school?

Meet with a partner and share your drawings and sentences. Talk with your partner about how your drawing matches the story.

Name ________________________________ Date ______________

Read the word problem.
Draw and label.
Write a number sentence and a statement that matches the story.

Shanika ate 7 mini-pretzels in the morning. She ate the rest of her mini-pretzels in the afternoon. She ate 13 mini-pretzels altogether that day. How many mini-pretzels did Shanika eat in the afternoon?

Read

Yesterday, I saw 11 birds on a branch. Three birds joined them on the branch. How many birds were on the branch then?

Draw

Write

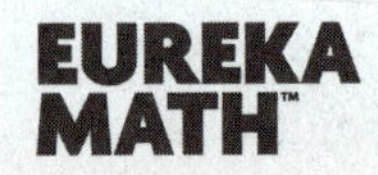

Name ______________________________ Date ______________

Read the word problem.
Draw and label.
Write a number sentence and a statement that match the story.

1. Jose sees 11 frogs on the shore. Some of the frogs hop into the water. Now, there are 8 frogs on the shore. How many frogs hopped into the water?

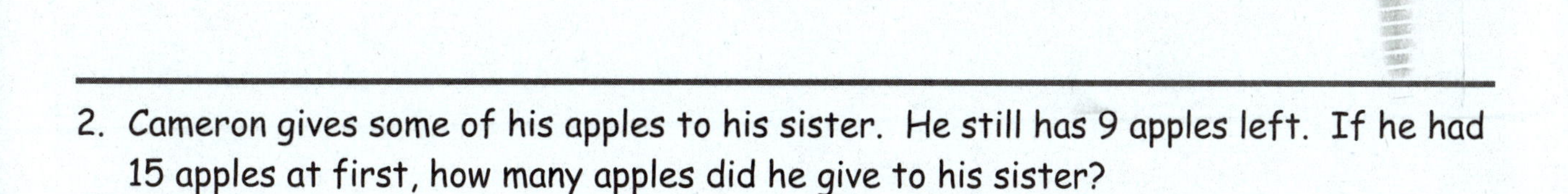

2. Cameron gives some of his apples to his sister. He still has 9 apples left. If he had 15 apples at first, how many apples did he give to his sister?

3. Molly had 16 books. She loaned some to Gia. How many books did Gia borrow if Molly has 8 books left?

4. Eighteen baby goats were playing outside. Some went into the barn. Nine stayed outside to play. How many baby goats went inside?

Meet with a partner and share your drawings and sentences. Talk with your partner about how your drawing tells the story.

Name _______________________________ Date _______________

Read the word problem.
Draw and label.
Write a number sentence and a statement that matches the story.

There were 18 dogs splashing in a puddle. Some dogs left. There are 9 dogs still splashing in the puddle. How many dogs are left?

Read

Micah had 16 trucks and lost 9 of them. Charles had 1 truck and received 6 more trucks from his mother. Who has more trucks, Micah or Charles?

Draw

Write

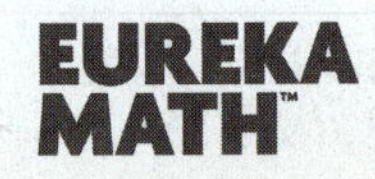

Name ______________________________ Date ______________

Use the expression cards to play Memory. Write the matching expressions to make true number sentences.

1. [ ] = [ ]

2. [ ] = [ ]

3. [ ] = [ ]

4. [ ] = [ ]

5. [ ] = [ ]

6. Write a true number sentence using the expressions that you have left over. Use pictures and words to show how you know two of the expressions have the same unknown numbers.

7. Use other facts you know to write at least two true number sentences similar to the type above.

8. The following addition number sentences are FALSE. Change one number in each problem to make a TRUE number sentence, and rewrite the number sentence.

a. 8 + 5 = 10 + 2 ____________________

b. 9 + 3 = 8 + 5 ____________________

c. 10 + 3 = 7 + 5 ____________________

9. The following subtraction number sentences are FALSE. Change one number in each problem to make a TRUE number sentence, and rewrite the number sentence.

a. 12 - 8 = 1 + 2 ____________________

b. 13 - 9 = 1 + 4 ____________________

c. 1 + 3 = 14 - 9 ____________________

Name ______________________________ Date ______________

You are given these new expression cards. Write matching expressions to make true number sentences.

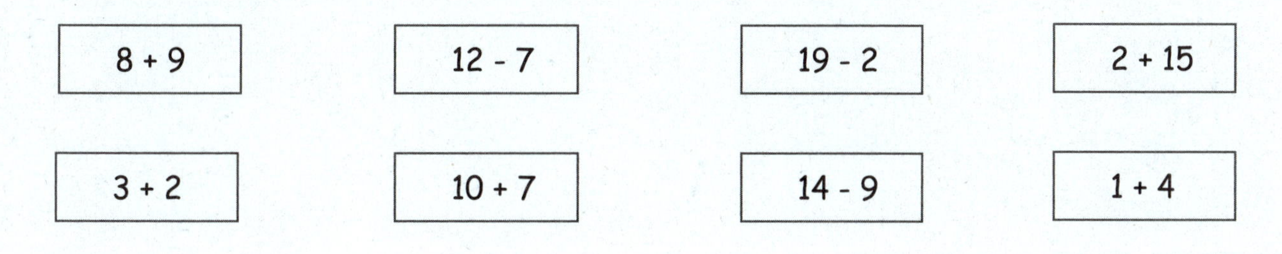

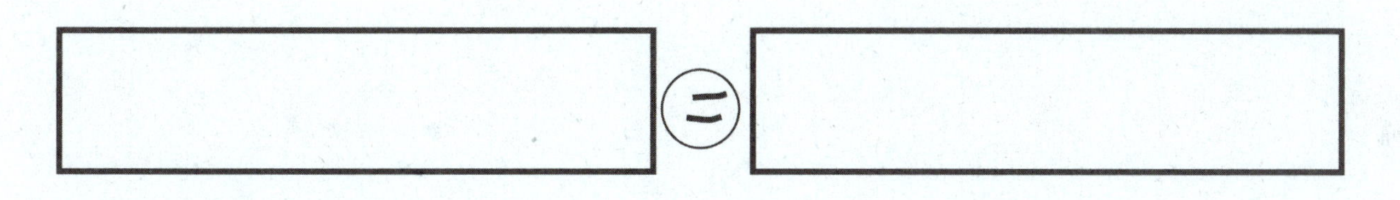

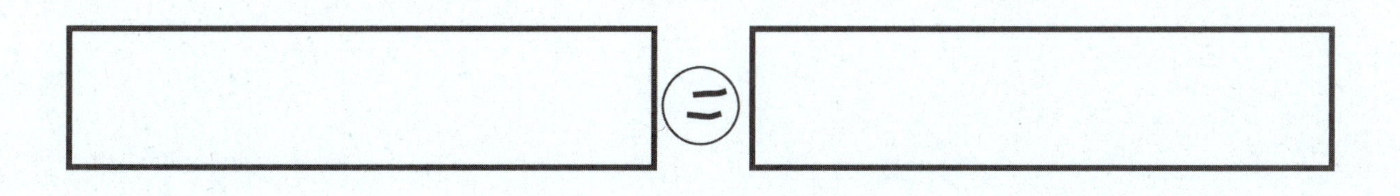

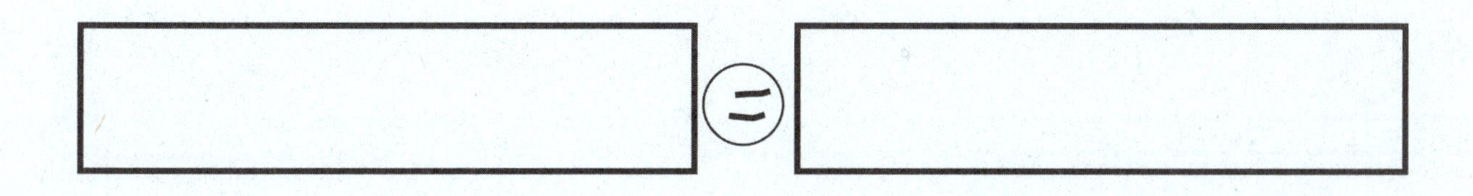

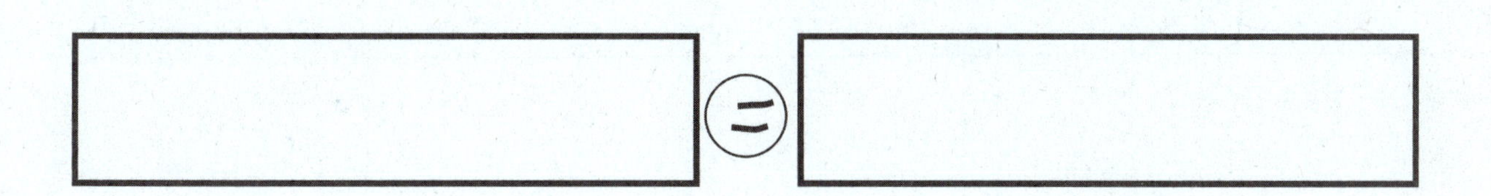

Read

Ruben has 18 toy cars. His car carrier holds 10 toy cars. If Ruben's carrier is full, how many cars are in the carrier, and how many cars are outside of the carrier?

Draw

Write

____

____

____

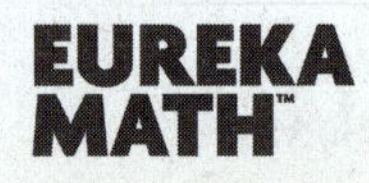

Show the total and tens and ones with Hide Zero cards.
Write how many **tens** and **ones**.

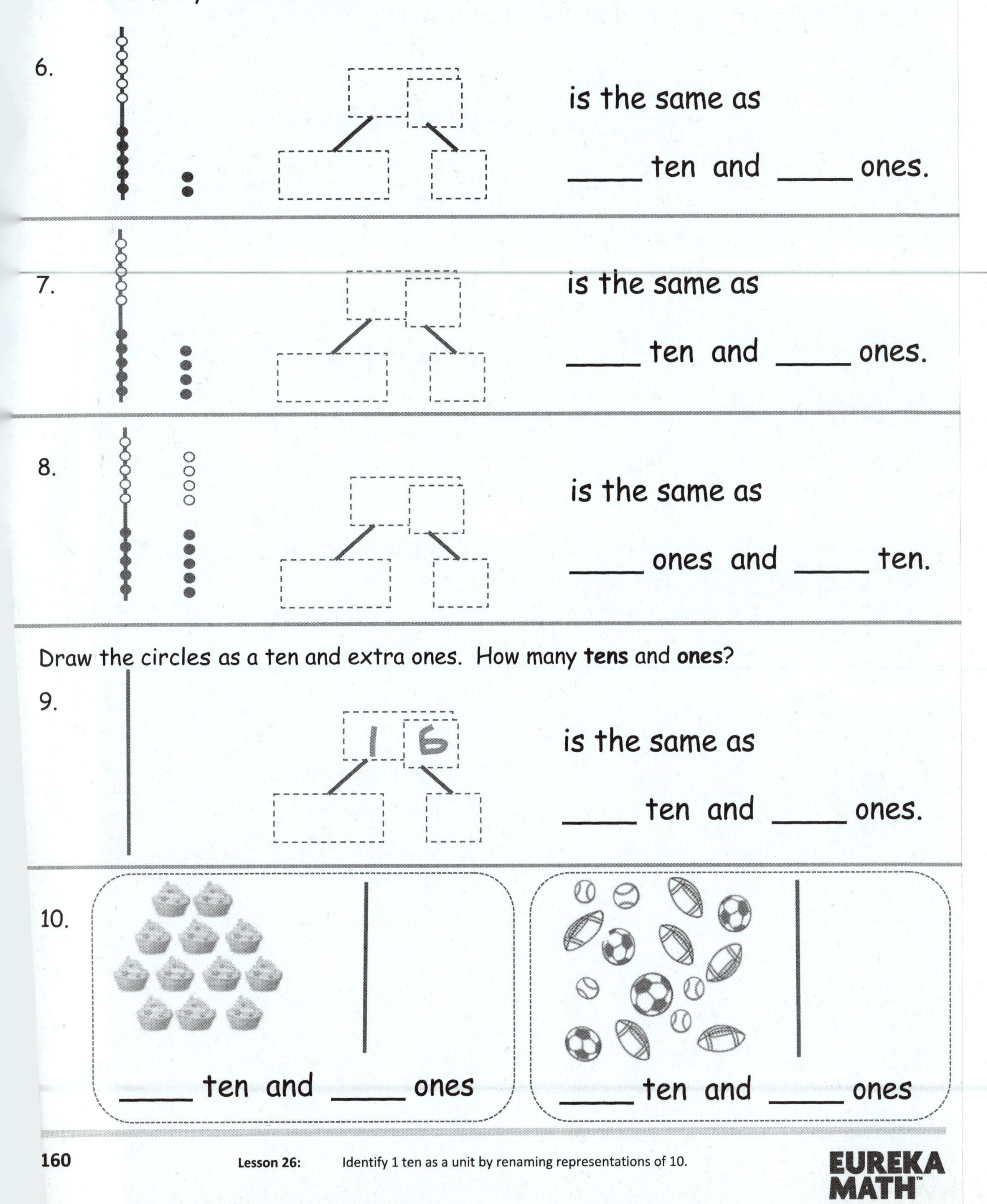

EUREKA MATH

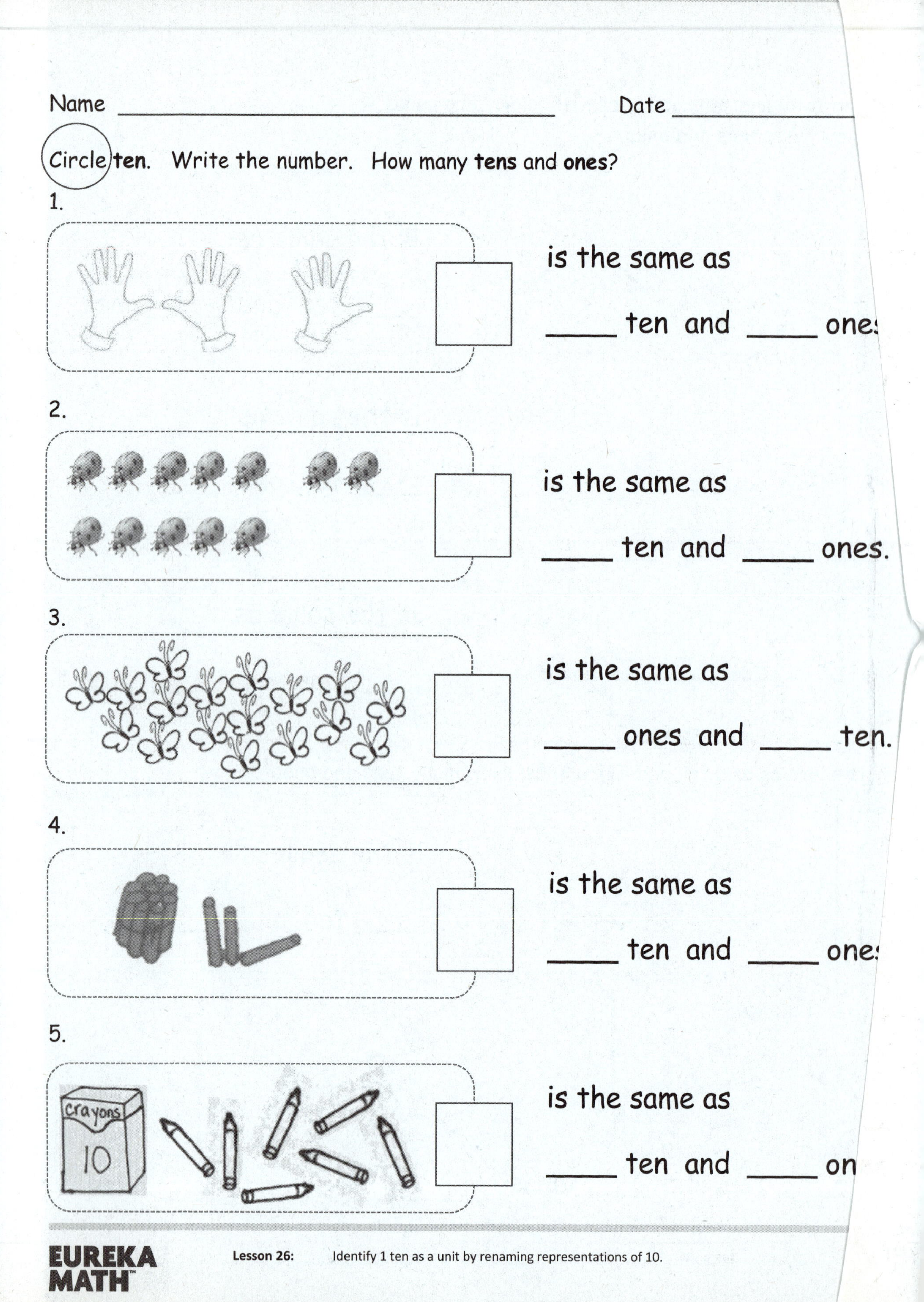
Name
Date
Circle ten. Write the number. How many tens and ones?
1.
is the same as
____ ten and ____ one
2.
is the same as
____ ten and ____ ones.
3.
is the same as
____ ones and ____ ten.
4.
is the same as
____ ten and ____ one
5.
crayons
10
is the same as
____ ten and ____ on

Name ______________________________ Date ________________

Match the pictures of tens and ones to the Hide Zero cards. How many tens and ones?

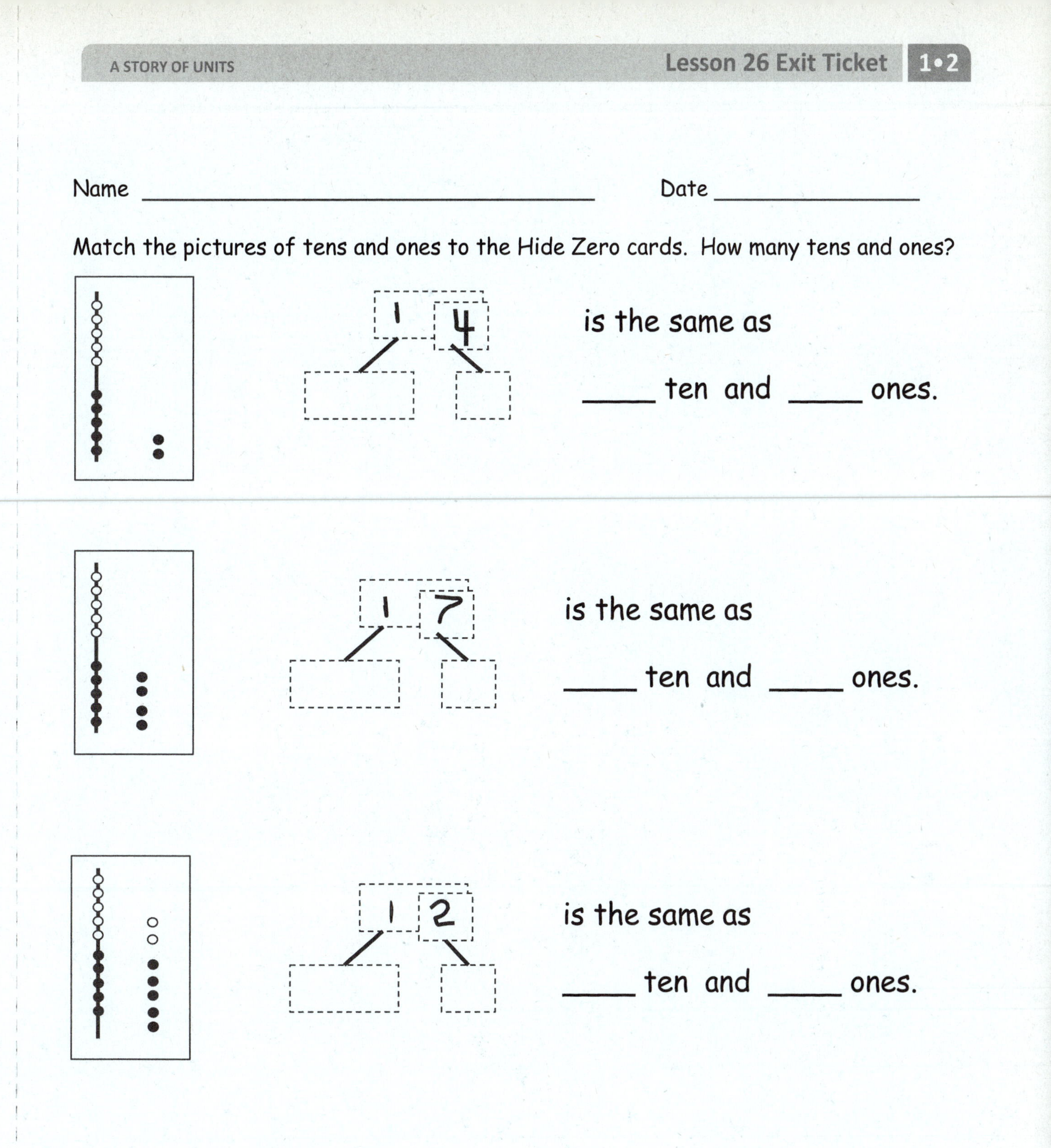

1 4 is the same as ____ ten and ____ ones.

1 7 is the same as ____ ten and ____ ones.

1 2 is the same as ____ ten and ____ ones.

Read

Ruben was putting away his 14 toy cars. He filled his car carrier and had 4 cars left that could not fit. How many cars fit in his car carrier?

Draw

Write

Name ______________________________ Date ______________

Solve the problems. Write your answers to show how many **tens** and **ones**. If there is only 1 ten, cross off the "s."

Add.

1. 12 + 6 = ☐☐

____ tens and ____ ones

2. 5 + 13 = ☐☐

____ tens and ____ ones

3. 8 + 7 = ☐☐

____ tens and ____ ones

4. ☐☐ = 8 + 12

____ tens and ____ ones

Subtract.

5. 17 - 4 = ☐☐

____ tens and ____ ones

6. 17 - 5 = ☐☐

____ tens and ____ ones

7. 14 - 6 = ☐☐

____ tens and ____ ones

8. ☐☐ = 16 - 7

____ tens and ____ ones

Read the word problem. **D**raw and label. **W**rite a number sentence and statement that matches the story. Rewrite your answer to show its tens and ones. If there is only 1 ten or 1 one, cross off the "s."

9. Frankie and Maya made 4 big sandcastles at the beach. If they made 10 small sandcastles, how many total sandcastles did they make?

____ tens and ____ ones

---

10. Ronnie has 8 stickers that are stars. Her friend Sina gives her 7 more. How many stickers does Ronnie have now?

____ tens and ____ ones

---

11. We tied 14 balloons to the tables for a party, but 3 floated away! How many balloons were still tied to the tables?

____ tens and ____ ones

---

12. I ate 5 of the 16 strawberries that I picked. How many did I have left over?

____ tens and ____ ones

Name ______________________________ Date ______________

Solve the problems. Write the answers to show how many tens and ones. If there is only one ten, cross off the "s."

1.

13 + 6 = [ ][ ]

____ tens and ____ ones

2.

7 + 6 = [ ][ ]

____ tens and ____ ones

Read the word problem. **D**raw and label. **W**rite a number sentence and statement that matches the story. Rewrite your answer to show its tens and ones.

3. Kendrick went bowling. He knocked down 16 pins in the first two frames. If he knocked down 9 in the first frame, how many pins did he knock down in the second frame?

____ tens and ____ ones

Read

Ruben has 7 blue cars and 6 red cars. If Ruben puts all of the blue cars in his car carrier that carries 10 cars, how many red cars will fit in the carrier, and how many will be left out of the carrier?

Draw

Write

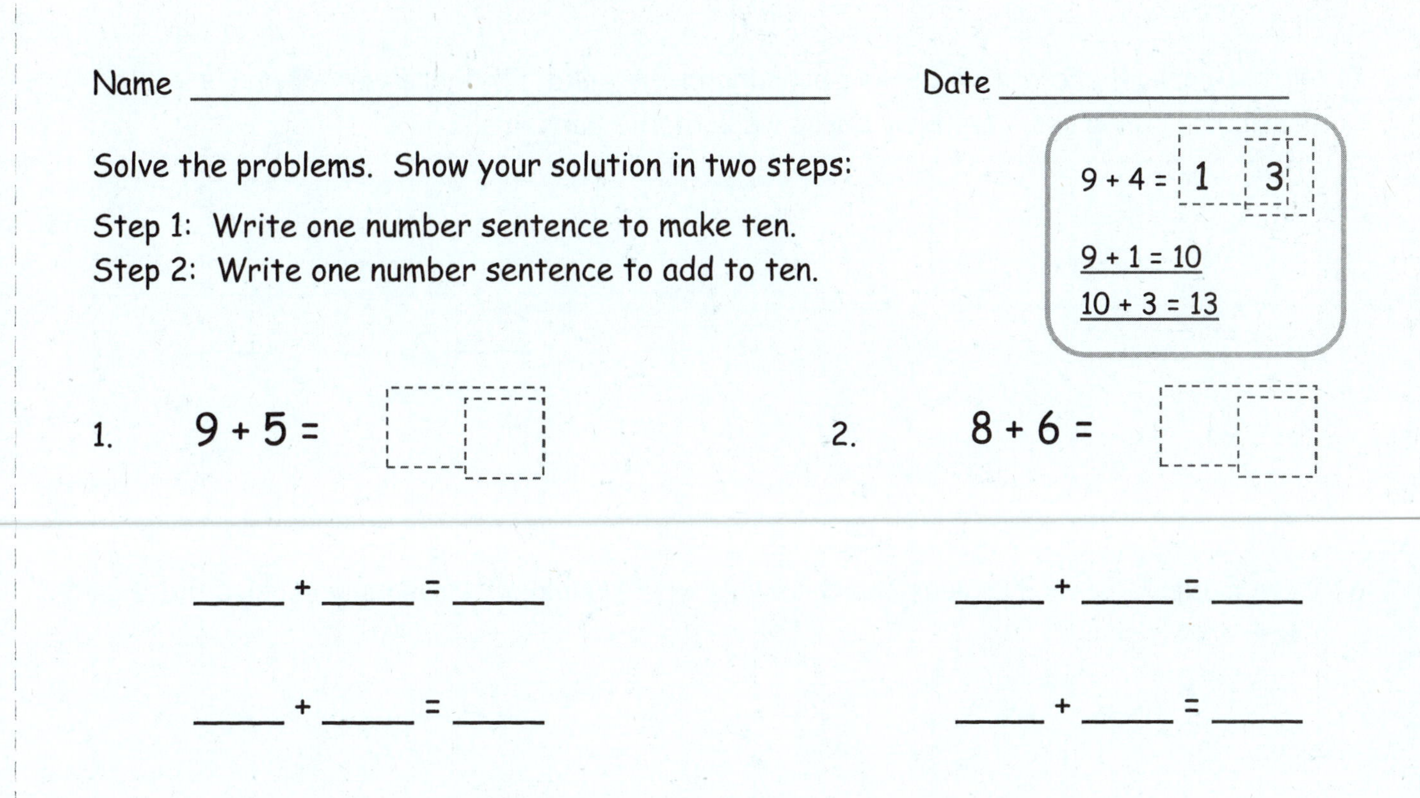

Name ______________________ Date ____________

Solve the problems. Show your solution in two steps:

Step 1: Write one number sentence to make ten.

Step 2: Write one number sentence to add to ten.

1. 9 + 5 = ☐☐

2. 8 + 6 = ☐☐

___ + ___ = ___

___ + ___ = ___

___ + ___ = ___

___ + ___ = ___

Solve. Then, write a statement to show your answer.

3. Su-Hean put together a collage with 9 pictures. Adele put together another collage with 6 pictures. How many pictures did they use?

___ + ___ = ___

___ + ___ = ___

4. Imran has 8 crayons in his pencil case and 7 crayons in his desk. How many crayons does Imran have altogether?

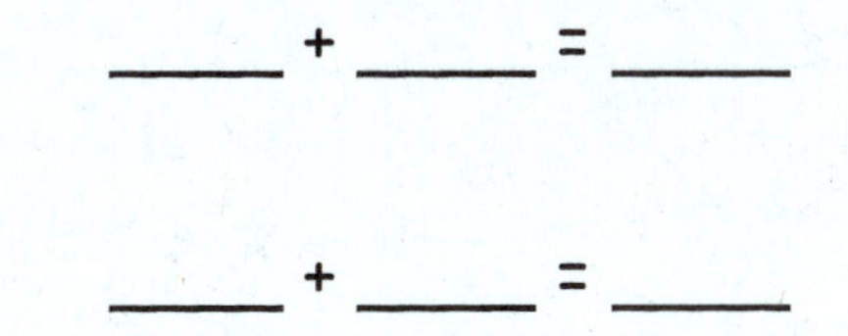

___ + ___ = ___

___ + ___ = ___

5. At the park, there were 4 ducks swimming in the pond. If there were 9 ducks resting on the grass, how many ducks were at the park in all?

_____ + _____ = _____

_____ + _____ = _____

6. Cece made 7 frosted cookies and 8 cookies with sprinkles. How many cookies did Cece make?

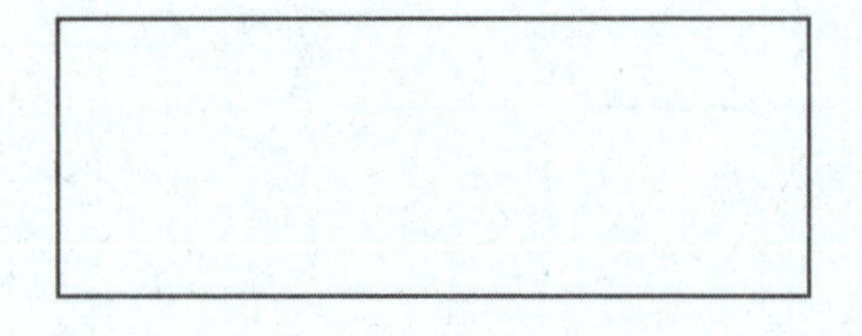

7. Payton read 8 books about dolphins and whales. She read 9 books about dogs and cats. How many books did she read about animals altogether?

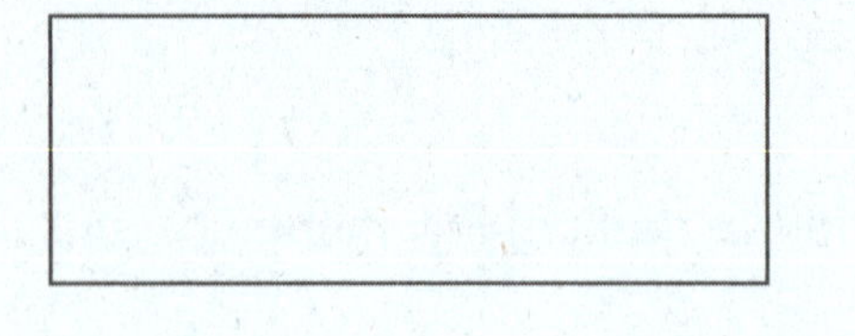

Name ______________________________ Date ______________

Solve the problems. Write your answers to show how many **tens** and **ones**.

9 + 7 = 1 6

9 + 1 = 10

10 + 6 = 16

1. 9 + 4 =

____ + ____ = ____

____ + ____ = ____

2. 8 + 7 =

____ + ____ = ____

____ + ____ = ____

Read

Hae Jung had 13 markers, and she gave some to Lily. If Hae Jung then had 5 markers, how many markers did she give to Lily?

Draw

Write

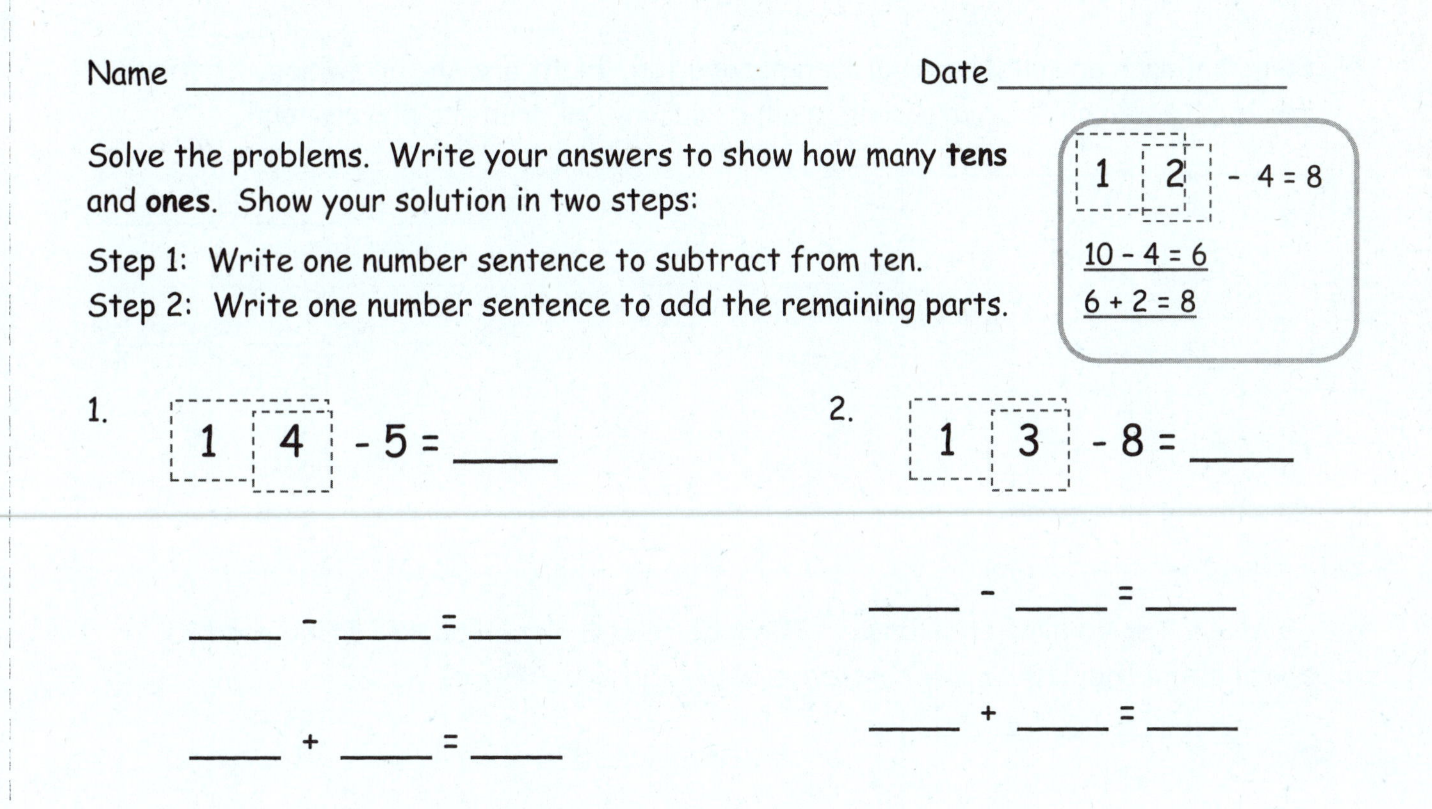

Name ______________________ Date ____________

Solve the problems. Write your answers to show how many **tens** and **ones**. Show your solution in two steps:

Step 1: Write one number sentence to subtract from ten.

Step 2: Write one number sentence to add the remaining parts.

1 2 - 4 = 8

10 - 4 = 6

6 + 2 = 8

1. 1 4 - 5 = ____

____ - ____ = ____

____ + ____ = ____

2. 1 3 - 8 = ____

____ - ____ = ____

____ + ____ = ____

3. Tatyana counted 14 frogs. She counted 8 swimming in the pond and the rest sitting on lily pads. How many frogs did she count sitting on lily pads?

14 - 8 = ____

____ - ____ = ____

____ + ____ = ____

4. This week, Maria ate 5 yellow plums and some red plums. If she ate 11 plums in all, how many red plums did Maria eat?

____ - ____ = ____

____ + ____ = ____

5. Some children are on the playground playing tag. Eight are on the swings. If there are 16 children on the playground in all, how many children are playing tag?

---

6. Oziah read some nonfiction books. Then, he read 6 fiction books. If he read 18 books altogether, how many nonfiction books did Oziah read?

---

7. Hadley has 9 buttons on her jacket. She has some more buttons on her shirt. Hadley has a total of 17 buttons on her jacket and shirt. How many buttons does she have on her shirt?

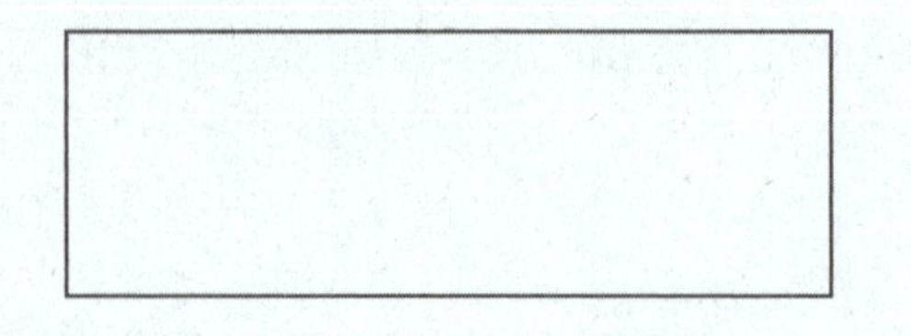

Name ______________________________ Date ______________

Solve the problems. Write your answers to show how many **tens** and **ones**.

1 2 - 5 = 7
10 - 5 = 5
5 + 2 = 7

1. 1 5 - 6 = ____

____ - ____ = ____

____ + ____ = ____

2. 1 4 - 8 = ____

____ - ____ = ____

____ + ____ = ____

Grade 1
Module 3

Read

Nigel and Corey each have new pencils that are the same length. Corey uses his pencil so much that he needs to sharpen it several times. Nigel doesn't use his at all. Nigel and Corey compare pencils. Whose pencil is longer? Draw a picture to show your thinking.

Draw

Write

Name ______________________________ Date ______________

Write the words **longer than** or **shorter than** to make the sentences true.

1.

Abby

Spot

Abby is ________________ Spot.

2.

A B

B is ________________A.

3.

The American flag hat

is ____________________________

the chef hat.

4.

The darker bat's wingspan

is ____________________________

the lighter bat's wingspan.

5.

B

A

Guitar B is

____

Guitar A.

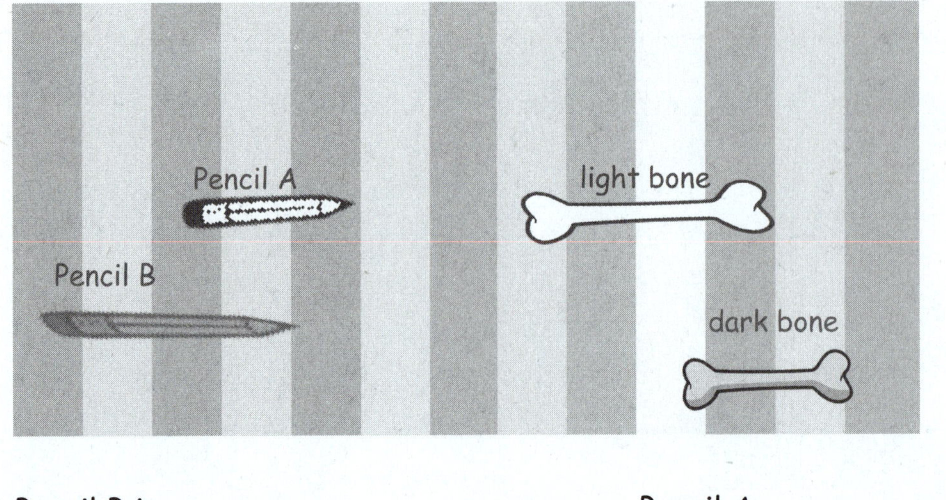

6. Pencil B is ______________________________ Pencil A.

7. The dark bone is ______________________________ the light bone.

8. Circle true or false.
   The light bone is shorter than Pencil A. **True** or **False**

---

9. Find 3 school supplies. Draw them here in order from **shortest** to **longest**.
   Label each school supply.

Name ______________________________ Date ______________

Write the words **longer than** or **shorter than** to make the sentence true.

A

B

Shoe A is ____________________ Shoe B.

Read

Jordan has 3 stuffed animals: a giraffe, a bear, and a monkey. The giraffe is taller than the monkey. The bear is shorter than the monkey. Sketch the animals from shortest to tallest to show how tall each animal is.

Draw

Write

Name ______________________________ Date ______________

1. Use the paper strip provided by your teacher to measure each **picture**. Circle the words you need to make the sentence true. Then, fill in the blank.

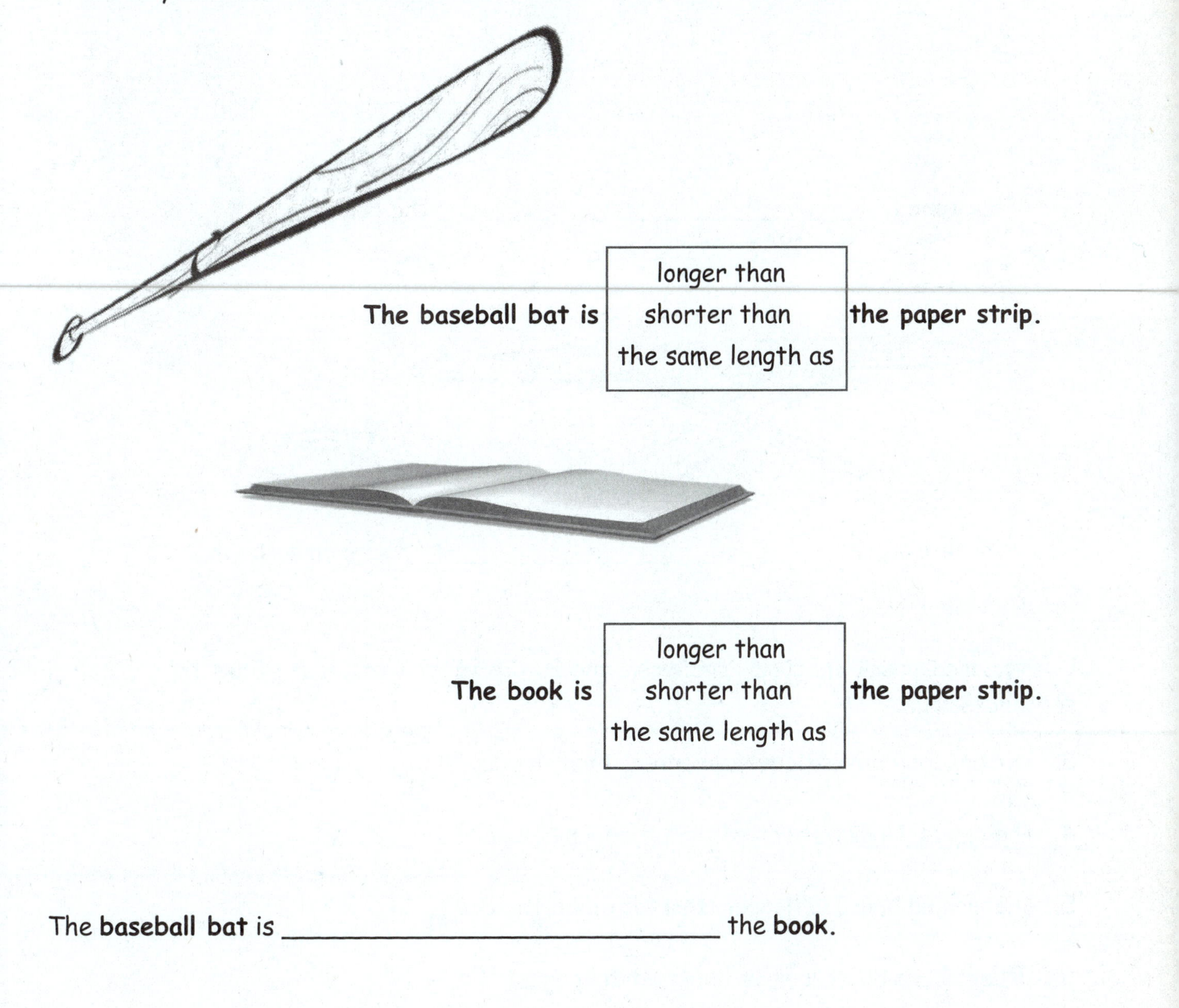

The baseball bat is

longer than
shorter than
the same length as

the paper strip.

The book is

longer than
shorter than
the same length as

the paper strip.

The **baseball bat** is ______________________________ the **book.**

2. Complete the sentences with **longer than**, **shorter than**, or **the same length as** to make the sentences true.

a.

The **tube** is ______________________ the **cup**.

b.

The **iron** is ______________________ the **ironing board**.

Use the measurements from Problems 1 and 2. Circle the word that makes the sentences true.

3. The baseball bat is (**longer/shorter**) than the cup.

4. The cup is (**longer/shorter**) than the ironing board.

5. The ironing board is (**longer/shorter**) than the book.

6. Order these objects from shortest to longest:

cup, tube, and paper strip

______________ ______________ ______________

Draw a picture to help you complete the measurement statements. Circle the words that make each statement true.

7. Sammy is taller than Dion.
   Janell is taller than Sammy.
   Dion is (**taller than/shorter than**) Janell.

8. Laura's necklace is longer than Mihal's necklace.
   Laura's necklace is shorter than Sarai's necklace.
   Sarai's necklace is (**longer than/shorter than**) Mihal's necklace.

Name ___________________________________ Date _______________

Draw a picture to help you complete the measurement statements. Circle the words that make each statement true.

Tanya's doll is shorter than Aline's doll.

Mira's doll is taller than Aline's doll.

Tanya's doll is (**taller than/shorter than**) Mira's doll.

If ________________ is longer than
(classroom object)

my foot and

________________ is shorter than my
(classroom object)

foot, then

________________ is longer than
(classroom object)

________________.
(classroom object)

My foot is about the same length as ________________.
(classroom object)

indirect comparison statements

Read

Draw one picture to match both of these sentences:

The book is longer than the index card. The book is shorter than the folder.

Which is longer, the index card or the folder? Write a statement comparing the two objects. Use your drawings to help you answer the question.

Draw

Write

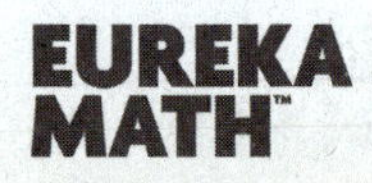

Name ______________________________ Date ______________

1. In a playroom, Lu Lu cut a piece of string that measured the distance from the doll house to the park. She took the same string and tried to measure the distance between the park and the store, but she ran out of string!

   Which is the longer path? Circle your answer.

the doll house to the park

the park to the store

Use the picture to answer the questions about the rectangles.

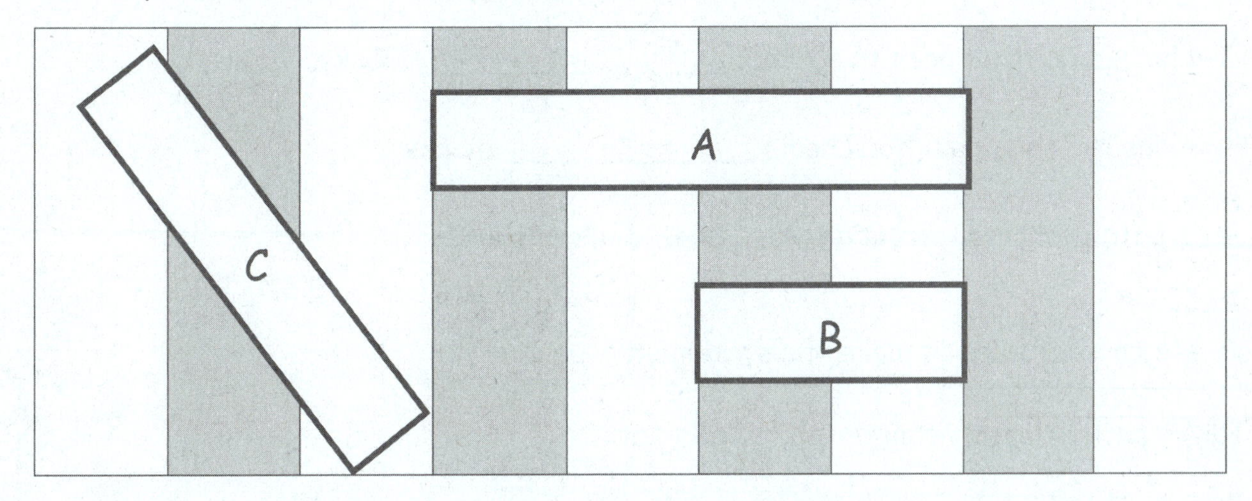

2. Which is the shortest rectangle? ________________

3. If Rectangle A is longer than Rectangle C, the longest rectangle is ____________.

4. Order the rectangles from shortest to longest:

______________ ______________ ______________

Use the picture to answer the questions about the students' paths to school.

Caitlyn's Path

Toby's Path

Joe's Path

School

5. How long is Caitlyn's path to school? ________________ blocks

6. How long is Toby's path to school? ______________ blocks

7. Joe's path is shorter than Caitlyn's. Draw Joe's path.

Circle the correct word to make the statement true.

8. Toby's path is **longer/shorter** than Joe's path.

9. Who took the shortest path to school? ______________________

10. Order the paths from shortest to longest.

_________________ _________________ _________________

Name ______________________________ Date ______________

Use the picture to answer the questions about the students' paths to the museum.

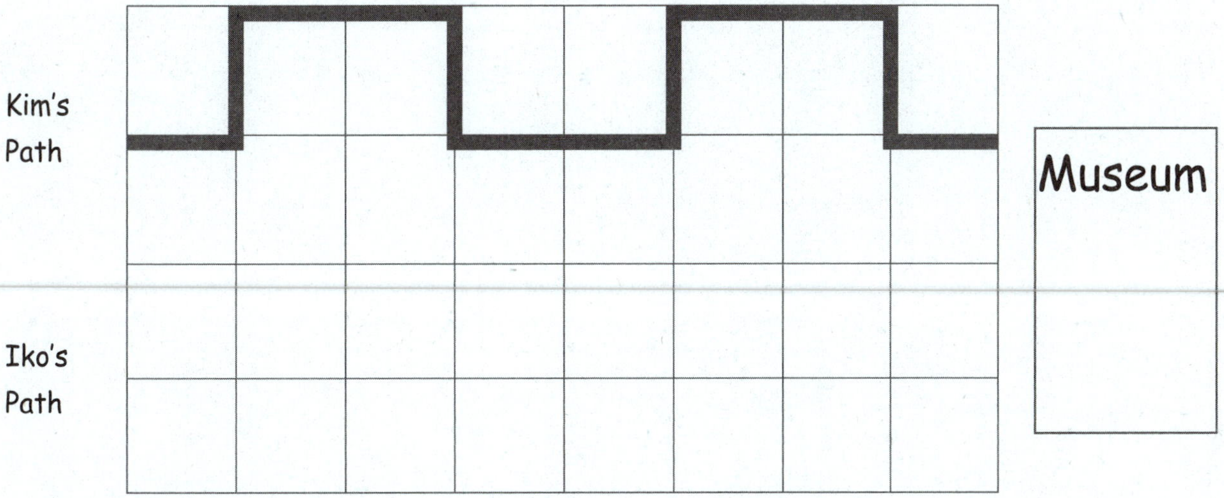

Kim's Path

Iko's Path

Museum

1. How long is Kim's path to the museum? ______________ blocks

2. Iko's path is shorter than Kim's path. Draw Iko's path.

Circle the correct word to make the statement true.

3. Kim's path is **longer/shorter** than Iko's path.

4. How long is Iko's path to the museum? ______________ blocks

Mary's House

Anne's House

Park

city blocks grid

Read

Joe ran a string from his room to his sister's room to measure the distance between them. When he tried to use the same string to measure the distance from his room to his brother's room, the string didn't reach! Which room was closer to Joe's room, his sister's or his brother's?

Draw

Write

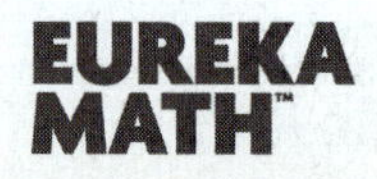

Name ______________________________ Date ______________

Measure the length of each picture with your cubes. Complete the statements below.

1. The pencil is ______ centimeter cubes long.

2. The pan is ______ centimeter cubes long.

3. The shoe is ______ centimeter cubes long.

4. The bottle is ______ centimeter cubes long.

5. The paintbrush is ______ centimeter cubes long.

6. The bag is ______ centimeter cubes long.

7. The ant is ______ centimeter cubes long.

8. The cupcake is ______ centimeter cubes long.

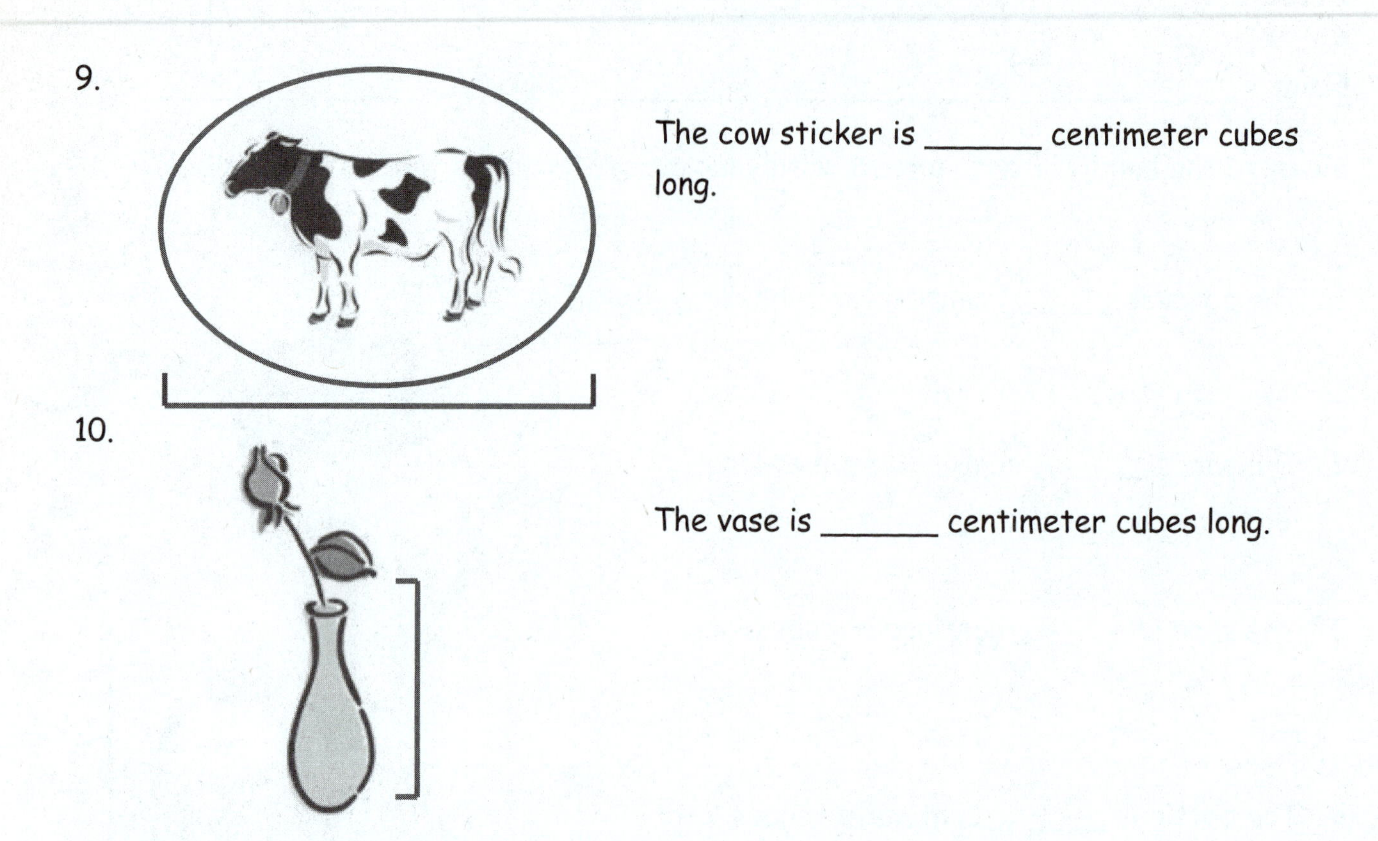

9. The cow sticker is ______ centimeter cubes long.

10. The vase is ______ centimeter cubes long.

11. Circle the picture that shows the correct way to measure.

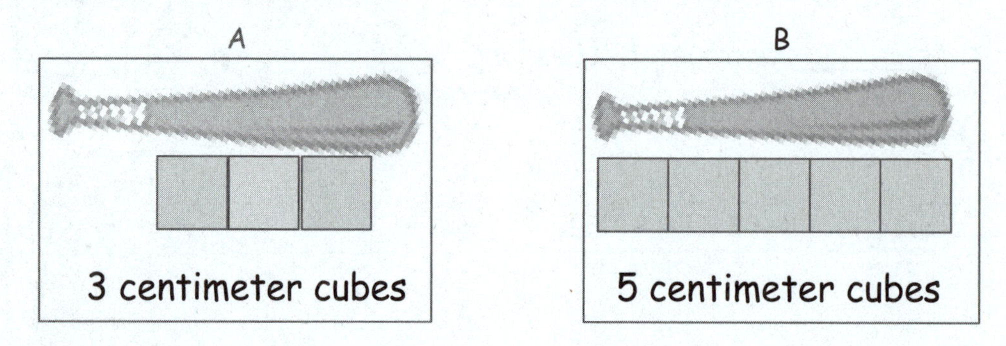

12. How would you fix the picture that shows an incorrect measurement?

____

____

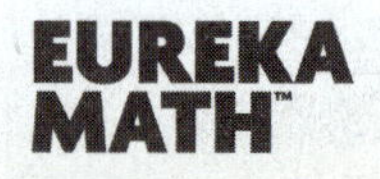

Name ____________________ Date ____________

1.

The picture frame is about ______ centimeter cubes long.

2.

The boy's *crutch* is about ______ centimeter cubes long.

Name ______________________________ Date ______________

Classroom Objects	Length Using Centimeter Cubes
glue stick	_____ centimeter cubes long
dry erase marker	_____ centimeter cubes long
craft stick	_____ centimeter cubes long
paper clip	_____ centimeter cubes long
	_____ centimeter cubes long
	_____ centimeter cubes long
	_____ centimeter cubes long

measurement recording sheet

Read

Amy used centimeter cubes to measure the length of her book.

She used 8 yellow centimeter cubes and 4 red centimeter cubes.

How many centimeter cubes long was her book?

Draw

Write

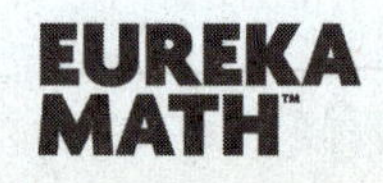

Name ______________________________ Date ______________

1. Circle the object(s) that are measured correctly.

a. 3 centimeters long

b. 5 centimeters long

c. 4 centimeters long

2. Measure the paper clip in 1(b) with your cubes. Then, check the cubes with your centimeter ruler.

The paper clip is _________ *centimeter cubes* long.

The paper clip is _________ *centimeters* long.

Be ready to explain why these are the same or different during the Debrief!

3. Use centimeter cubes to measure the length of each picture from left to right. Complete the statement about the length of each picture in centimeters.

a. The hamburger picture is _________ centimeters long.

b. The hot dog picture is _________ centimeters long.

c. The bread picture is __________ centimeters long.

4. Use centimeter cubes to measure the objects below. Fill in the length of each object.

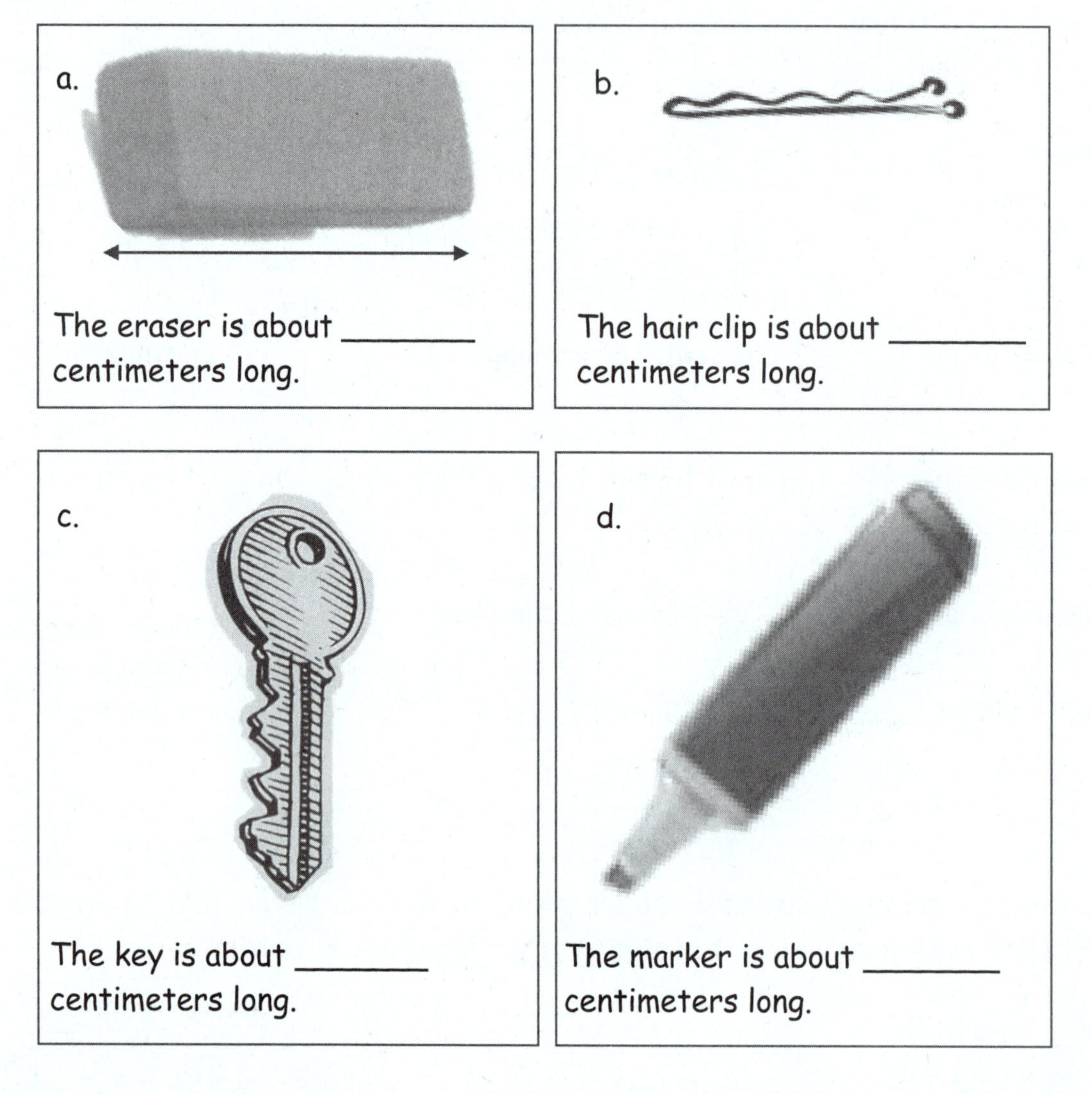

a. The eraser is about _______ centimeters long.

b. The hair clip is about _______ centimeters long.

c. The key is about _______ centimeters long.

d. The marker is about _______ centimeters long.

5. The eraser is longer than the __________________, but it is shorter than the __________________.

6. Circle the word that makes the sentence true.

   If a paper clip is shorter than the key, then the marker is **longer/shorter** than the paper clip.

Name ______________________________ Date ______________

Use the centimeter cubes to measure the items. Complete the sentences.

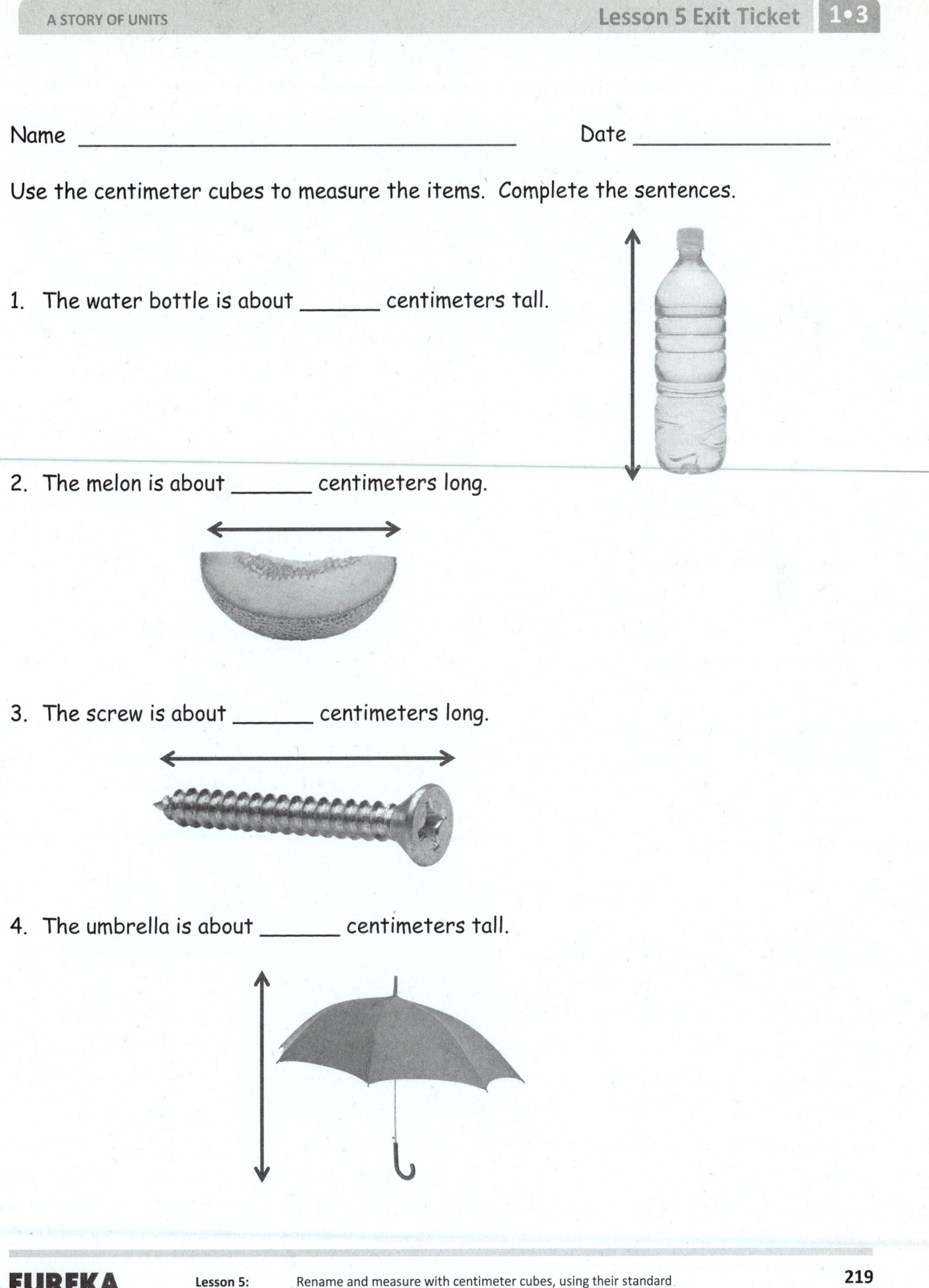

1. The water bottle is about ______ centimeters tall.

2. The melon is about ______ centimeters long.

3. The screw is about ______ centimeters long.

4. The umbrella is about ______ centimeters tall.

Read

Julia's lollipop is 15 centimeters long. She measured the lollipop with 9 red centimeter cubes and some blue centimeter cubes. How many blue centimeter cubes did she use? Remember to use the RDW process.

Draw

Write

Name ______________________________ Date ______________

1. Order the bugs from longest to shortest by writing the bug names on the lines. Use centimeter cubes to check your answer. Write the length of each bug in the space to the right of the pictures.

   The bugs from longest to shortest are

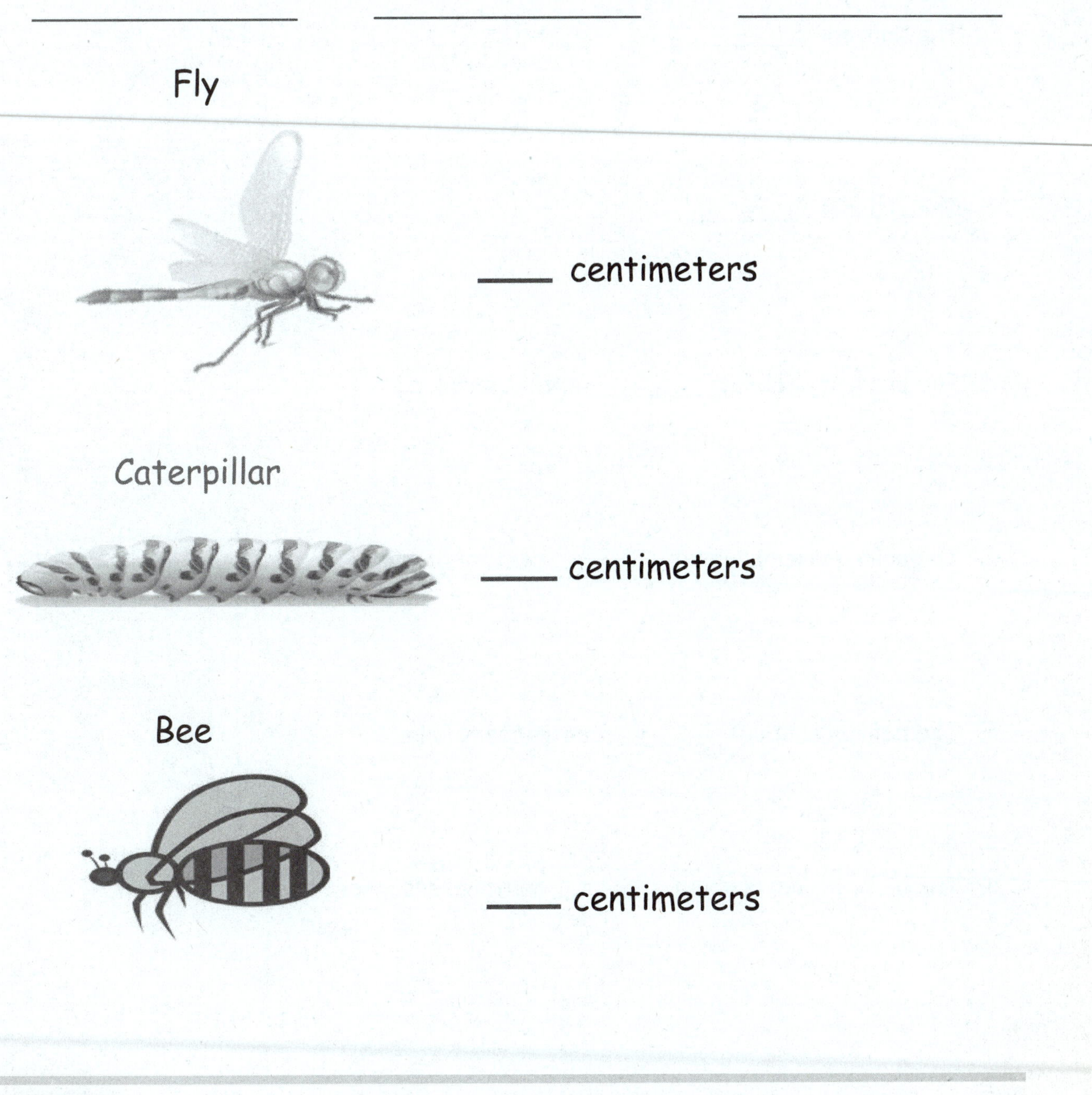

2. Order the objects below from shortest to longest using the numbers 1, 2, and 3. Use your centimeter cubes to check your answers, and then complete the sentences for problems d, e, f, and g.

   a. The noise maker: _______

   b. The balloon: _______

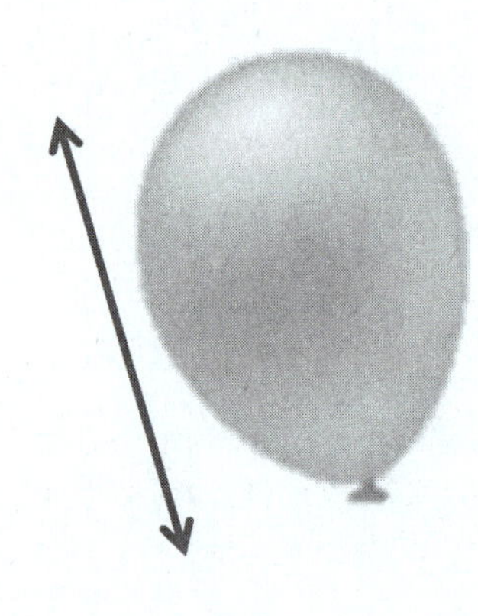

   c. The present: _______

   d. The present is about _______ centimeters long.

   e. The noise maker is about _______ centimeters long.

   f. The balloon is about _______ centimeters long.

   g. The noise maker is about _______ centimeters longer than the present.

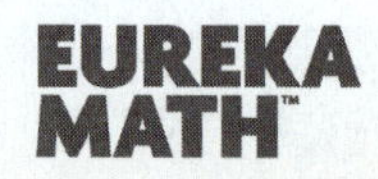

Use your centimeter cubes to model each length, and answer the question. Write a statement for your answer.

3. Peter's toy T. rex is 11 centimeters tall, and his toy Velociraptor is 6 centimeters tall. How much taller is the T. rex than the Velociraptor?

4. Miguel's pencil rolled 17 centimeters, and Sonya's pencil rolled 9 centimeters. How much less did Sonya's pencil roll than Miguel's?

5. Tania makes a cube tower that is 3 centimeters taller than Vince's tower. If Vince's tower is 9 centimeters tall, how tall is Tania's tower?

Name ______________________________ Date ____________

Read the measurements of the tool pictures.

The wrench is 8 centimeters long.

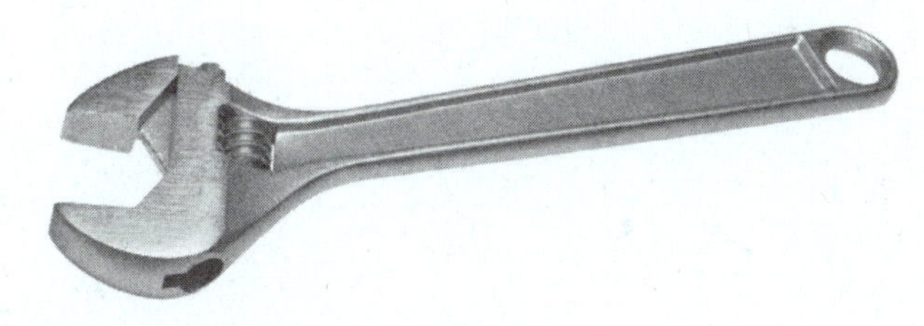

The screwdriver is 12 centimeters long.

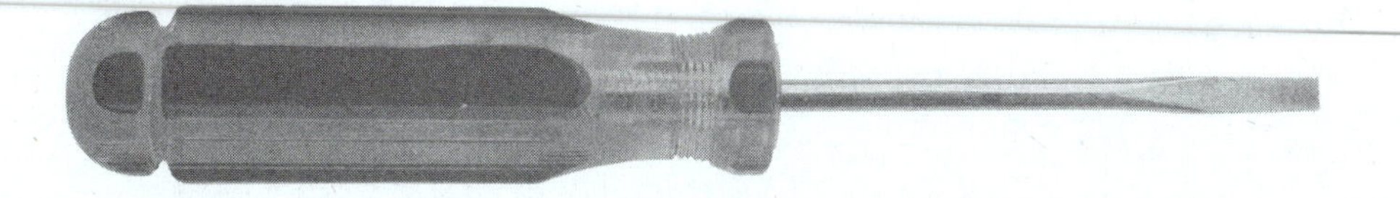

The hammer is 9 centimeters long.

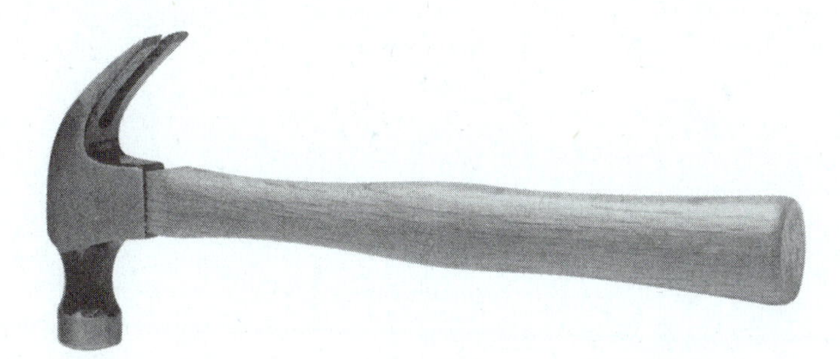

1. Order the pictures of the tools from shortest to longest.

_______________ _______________ _______________

2. How much longer is the screwdriver than the wrench?

The screwdriver is ______ centimeters longer than the wrench.

Read

When Corey measures his new pencil, he uses 19 centimeter cubes. After he sharpens the pencil, he needs 4 fewer centimeter cubes. How long is Corey's pencil after he sharpens it? Use centimeter cubes to solve the problem. Write a number sentence and a statement to answer the question.

Draw

Write

Name ______________________________ Date ________________

1. Measure the length of each object with **LARGE** paper clips. Fill in the chart with your measurements.

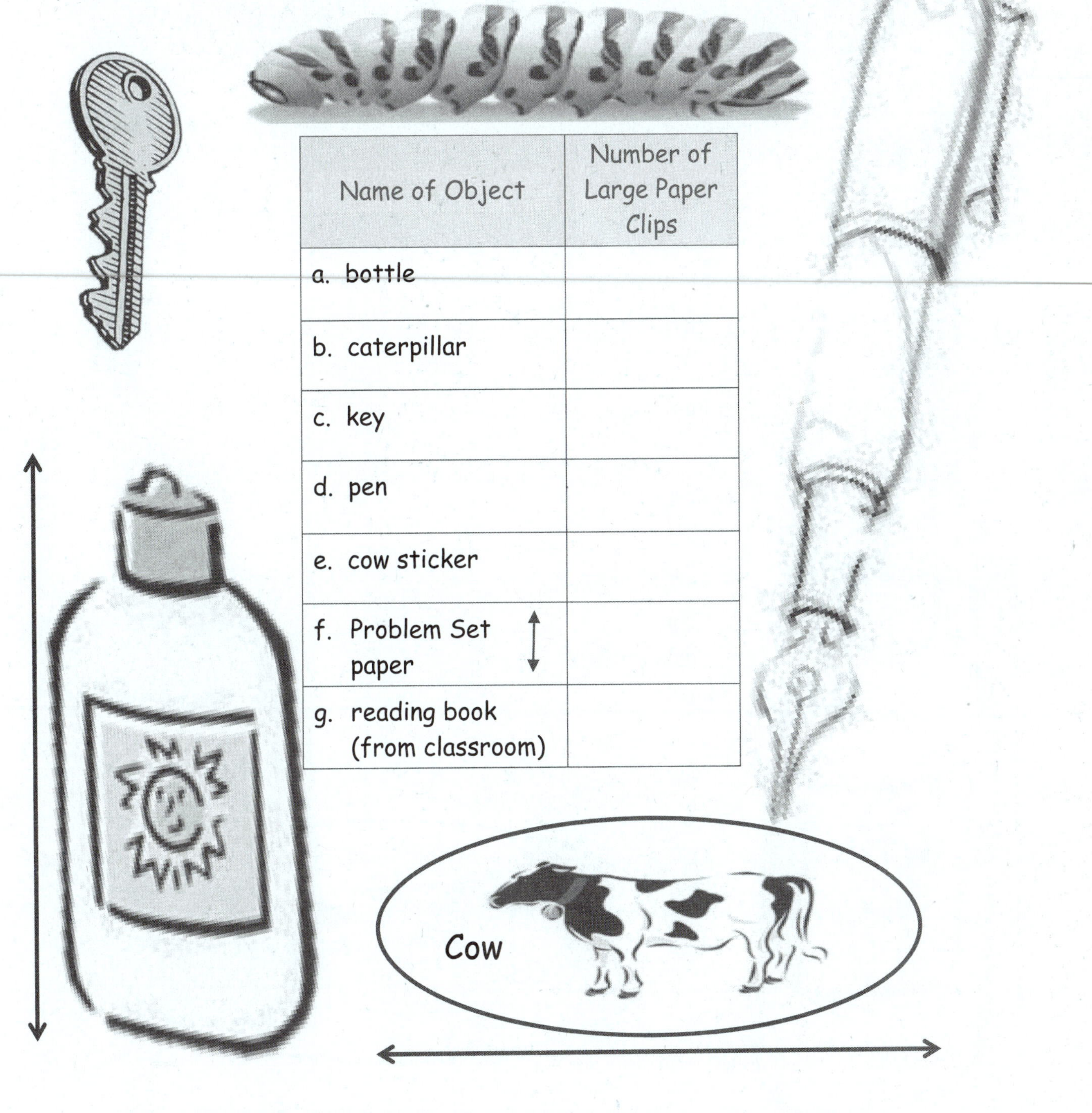

Name of Object	Number of Large Paper Clips
a. bottle	
b. caterpillar	
c. key	
d. pen	
e. cow sticker	
f. Problem Set paper ↕	
g. reading book (from classroom)	

2. Measure the length of each object with **SMALL** paper clips. Fill in the chart with your measurements.

Name of Object	Number of Small Paper Clips
a. bottle	
b. caterpillar	
c. key	
d. pen	
e. cow sticker	
f. Problem Set paper ↕	
g. reading book (from classroom)	

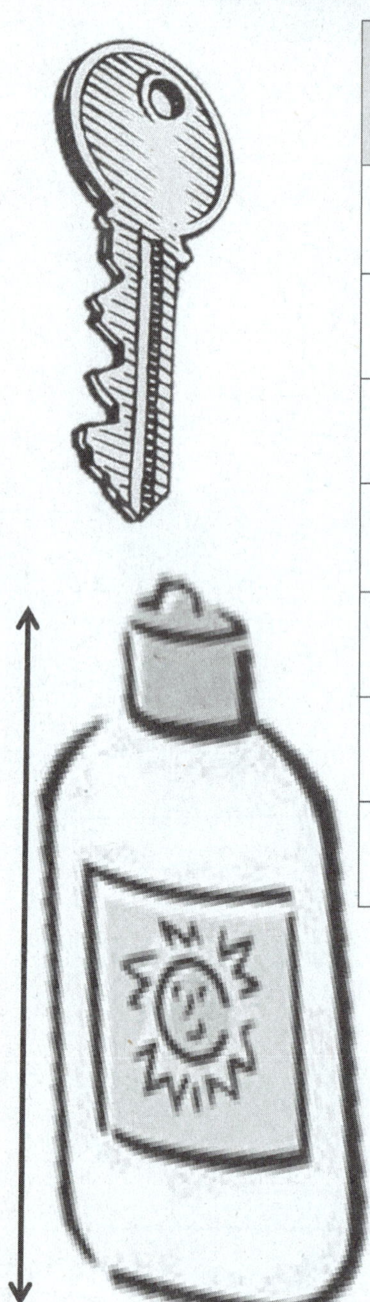

Cow

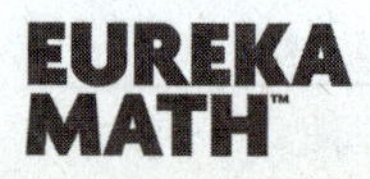

Name ______________________________ Date ________________

Measure the length of each object with **large** paper clips. Then, measure the length of each object with **small** paper clips. Fill in the chart with your measurements.

Name of Object	Number of Large Paper Clips	Number of Small Paper Clips
a. bow		
b. candle		
c. vase and flowers		

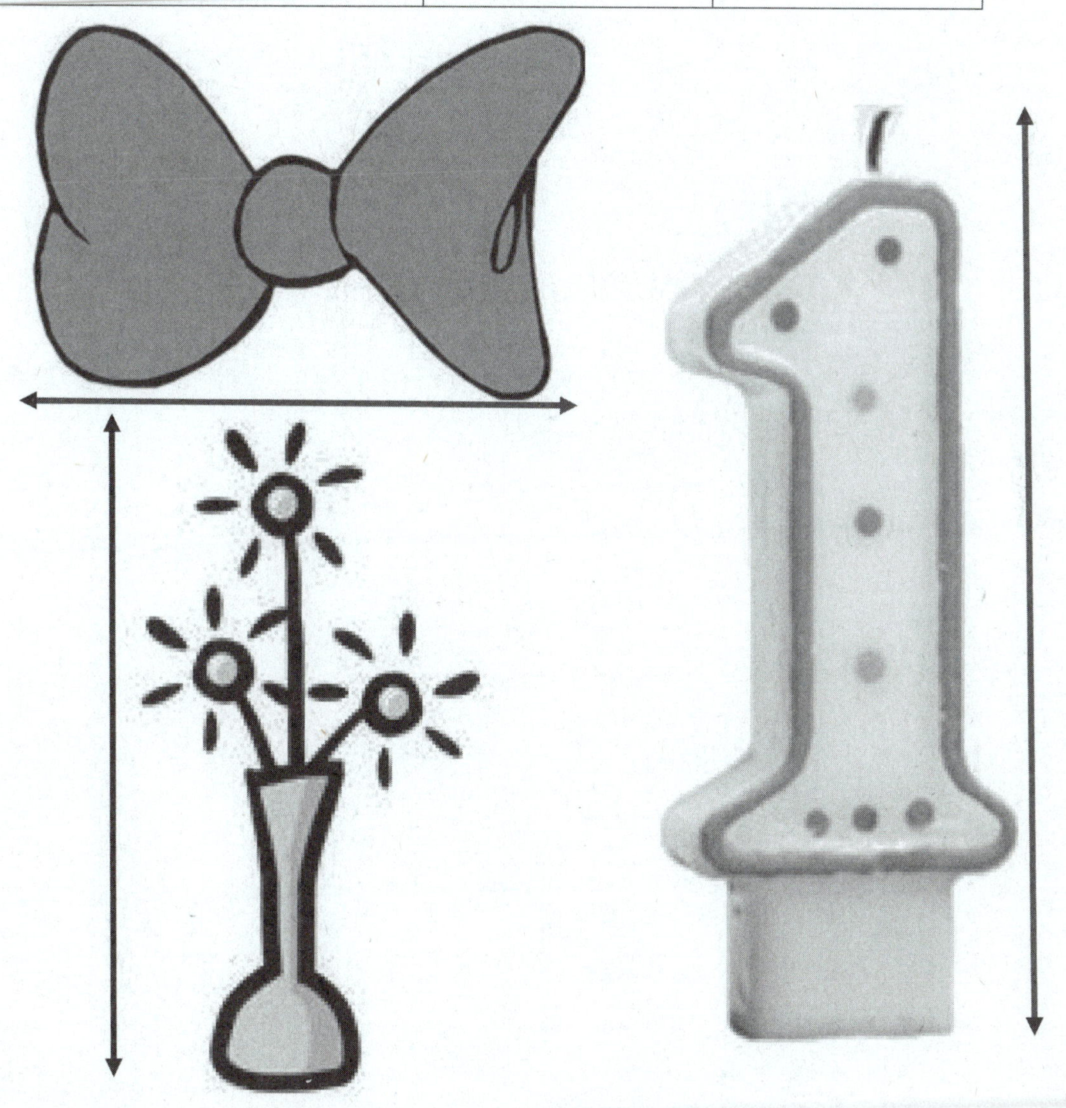

Read

I have 2 crayons. Each crayon is 9 centimeter cubes long. I also have a paintbrush. The paintbrush is the same length as 2 crayons. How many centimeter cubes long is the paintbrush? Use centimeter cubes to solve the problem. Then, draw a picture, and write a number sentence and a statement to answer the question.

Draw

Write

Name ______________________________ Date ________________

Circle the length unit you will use to measure. Use the same length unit for all objects.

Small Paper Clips

Large Paper Clips

Toothpicks

Centimeter Cubes

Measure each object listed on the chart, and record the measurement. Add the names of other objects in the classroom, and record their measurements.

Classroom Object	Measurement
a. glue stick	
b. dry erase marker	
c. unsharpened pencil	
d. personal white board	
e.	
f.	
g.	

Name ______________________________ Date ________________

Circle the length unit you will use to measure. Use the same length unit for all objects.

Small Paper Clips

Large Paper Clips

Toothpicks

Centimeter Cubes

Choose two objects in your desk that you would like to measure. Measure each object, and record the measurement.

Classroom Object	Measurement
a.	
b.	

Read

Corey buys a super-cool, extra-long crayon that is 14 centimeters long. His regular crayon is 9 centimeters long. Use centimeter cubes to find out how much longer Corey's new crayon is than his regular crayon.

Write a statement to answer the question. Write a number sentence to show what you did.

Draw

Write

Name ______________________________ Date ________________

1. Look at the picture below. How much **longer** is Guitar A than Guitar B?

GuitarA is ______ unit(s) **longer** than GuitarB.

2. Measure each object with centimeter cubes.

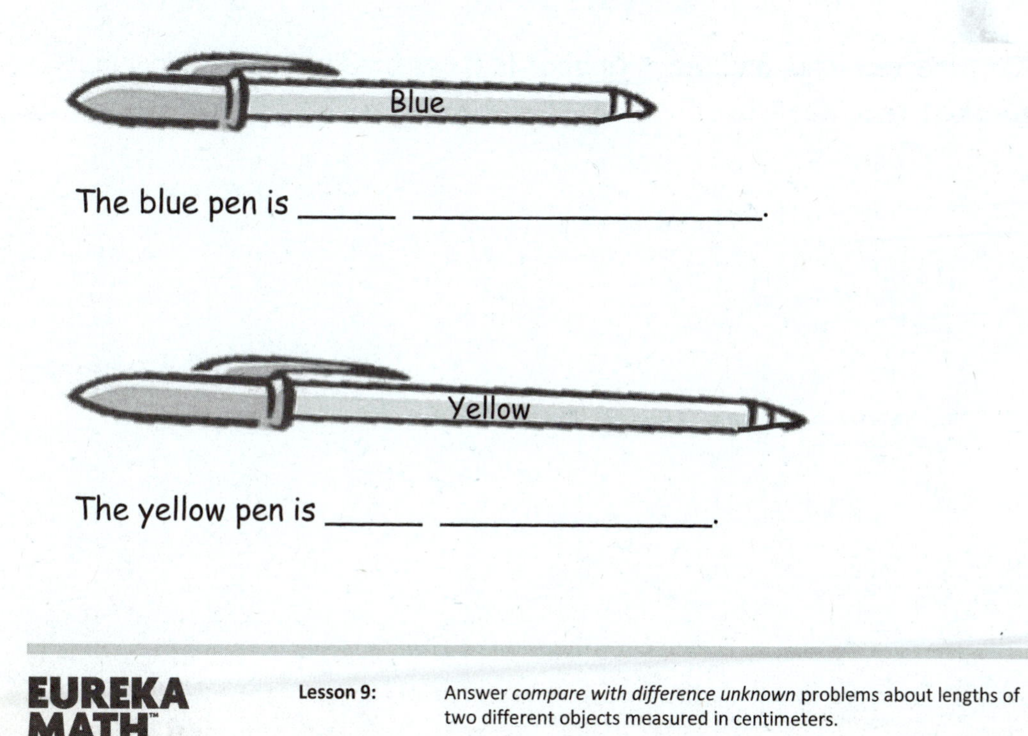

The blue pen is _____ ________________.

The yellow pen is _____ ______________.

3. How much **longer** is the yellow pen than the blue pen?

   The yellow pen is _____ centimeters **longer** than the blue pen.

4. How much **shorter** is the blue pen than the yellow pen?

   The blue pen is _____ centimeters **shorter** than the yellow pen.

Use your centimeter cubes to model each problem. Then, solve by drawing a picture of your model and writing a number sentence and a statement.

5. Austin wants to make a train that is 13 centimeter cubes long. If his train is already 9 centimeter cubes long, how many **more** cubes does he need?

6. Kea's boat is 12 centimeters long, and Megan's boat is 8 centimeters long. How much **shorter** is Megan's boat than Kea's boat?

7. Kim cuts a piece of ribbon for her mom that is 14 centimeters long. Her mom says the ribbon is 8 centimeters too long. How **long** should the ribbon be?

8. The tail of Lee's dog is 15 centimeters long. If the tail of Kit's dog is 9 centimeters long, how much **longer** is the tail of Lee's dog than the tail of Kit's dog?

Name ______________________________ Date ________________

Use your centimeter cubes to model the problem. Then, draw a picture of your model.

Mona's hair grew 7 centimeters. Claire's hair grew 15 centimeters. How much **less** did Mona's hair grow than Claire's hair?

Read

There were 14 items on the table to measure. I already measured 5 of them. How many more items are there to measure?

Draw

Write

There are ▢ more items to measure.

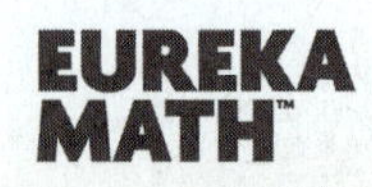

Name ______________________________ Date ________________

A group of people were asked to say their favorite color. Organize the data using tally marks, and answer the questions.

Red	
Green	
Blue	

1. How many people chose red as their favorite color? __________ people like red.

2. How many people chose blue as their favorite color? __________ people like blue.

3. How many people chose green as their favorite color? __________ people like green.

4. Which color received the least amount of votes? ______________

5. Write a number sentence that tells the total number of people who were asked their favorite color.

____

Name ____________________ Date __________

A group of students were asked what they ate for lunch. Use the data below to answer the following questions.

Student Lunches

Lunch	Number of Students
sandwich	3
salad	5
pizza	4

1. What is the **total** number of students who ate pizza? _____ student(s)

2. Which lunch was eaten by the **greatest** number of students? ______________

3. What is the total number of students who ate pizza or a sandwich?

   _____ student(s)

4. Write an addition sentence for the **total** number of students who were asked what they ate for lunch.

   ____

Read

Larry asked his friends whether dogs or cats are smarter. 9 of his friends think dogs are smarter, and 6 think cats are smarter. Make a table to show Larry's data collection. How many friends did he ask?

Draw

Write

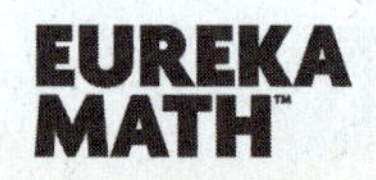

Name ______________________________ Date ________________

Welcome to Data Day! Follow the directions to **collect** and **organize** data. Then, **ask** and **answer questions** about the data.

- Choose a question. Circle your choice.
- Pick 3 answer choices.
- Ask your classmates the question, and show them the 3 choices. Record the data on a class list.
- Organize the data in the chart below.

Which fruit do you like best?	Which snack do you like best?	What do you like to do on the playground the most?	Which school subject do you like the best?	Which animal would you most like to be?

Answer Choices	Number of Students

- Complete the question sentence frames to ask questions about your data.
- Trade papers with a partner, and have your partner answer your questions.

1. How many students liked ______________ the best?

2. Which category received the fewest votes? ______________

3. How many more students liked ______________ than ________________?

4. What is the total number of students who liked ______________ or ______________ the best?

5. How many students answered the question? How do you know?

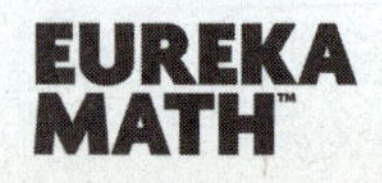

Name ______________________________ Date ________________

A class collected the information in the chart below. Students asked each other: Among stuffed animals, toy cars, and blocks, which is your favorite toy?

Then, they organized the information in this chart.

Toy	Number of Students
Stuffed Animals	11
Toy Cars	5
Blocks	13

1. How many students chose toy cars? __________

2. How many more students chose blocks than stuffed animals? __________

3. How many students would need to choose toy cars to equal the number of students who chose blocks? __________

Read

Kingston's class took a trip to the zoo. He collected data about his favorite African animals. He saw 2 lions, 11 gorillas, and 7 zebras. What might his table look like? Write one question your classmate can answer by looking at the table.

Draw

Write

Name ______________________ Date ____________

Use squares with no gaps or overlaps to organize the data from the picture. Line up your **squares** carefully.

Favorite Ice Cream Flavor □ = 1 student

Number of Students

Flavors	
□ vanilla	
■ chocolate	

1. How many **more** students liked chocolate than liked vanilla? ________ students

2. How many **total** students were asked about their favorite ice cream flavor?

________ students

Ties on Shoes Number of Students □ = 1 student

Types of Shoe Ties	
Velcro	□□□□□
laces	□□□□□□□□
no ties	□□□□□□□

3. Write a number sentence to show how many **total** students were asked about their shoes.

____

4. Write a number sentence to show how many **fewer** students have Velcro on their shoes than laces.

____

Each student in the class added a sticky note to show his or her favorite kind of pet. Use the graph to answer the questions.

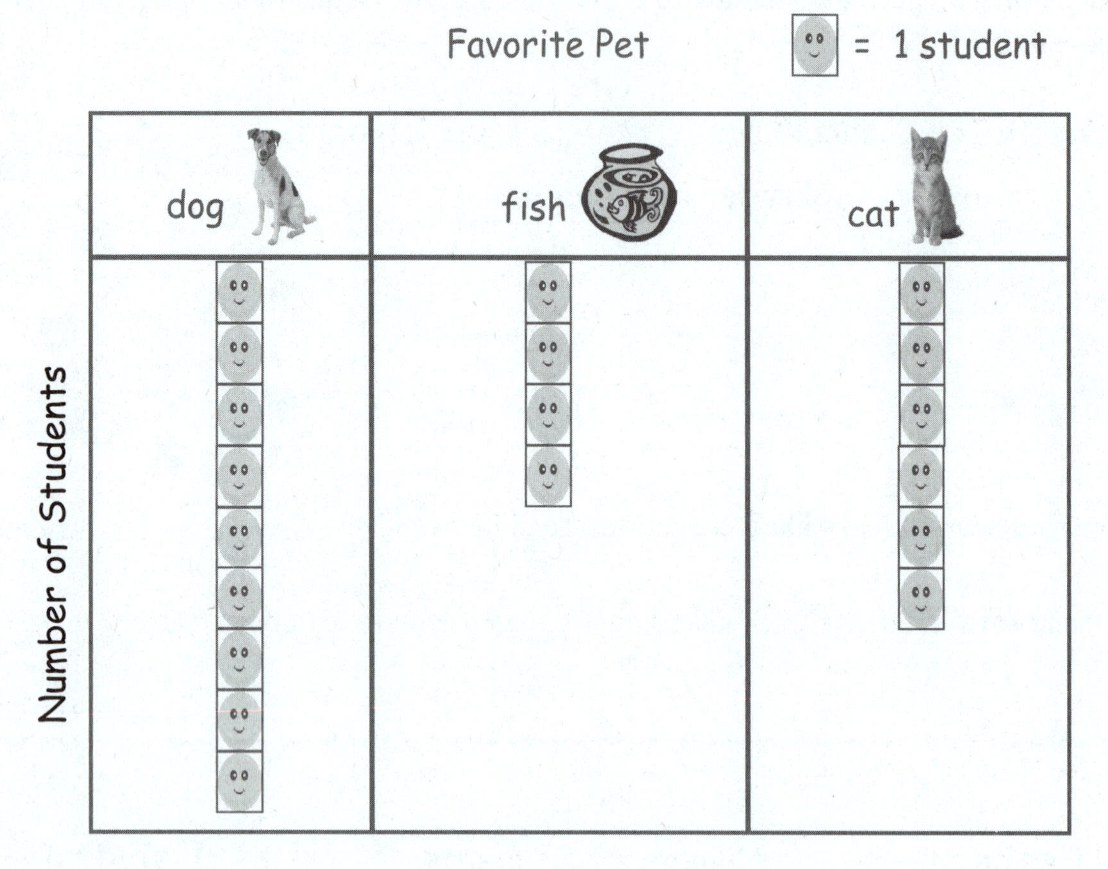

5. How many students chose dogs or cats as their favorite pet?

_________ students

6. How many more students chose dogs as their favorite pet than cats?

_________ students

7. How many more students chose cats than fish?

_________ students

Name ______________________________ Date ________________

Use squares with no gaps or overlaps to organize the data from the pictures.
Line up your **squares** carefully.

Favorite Animals at the Zoo

Zoo Animals	Number of Students
giraffe	
elephant	
lion	

Each picture represents 1 student's vote.

1. Write a number sentence to show how many **total** students were asked about their favorite animal at the zoo.

   ____

2. Write a number sentence to show how many **fewer** students like elephants than like giraffes.

   ____

Read

Zoe made friendship necklaces for her 3 closest friends. Make a graph to show the two colors of beads she used. She used 8 green beads for Lily, 4 purple beads for Jamilah, and 12 green beads for Sage. How many green beads did she use?

Draw

Write

Name ______________________________ Date ______________

Use the graph to answer the questions. Fill in the blank, and write a number sentence to the right to solve the problem.

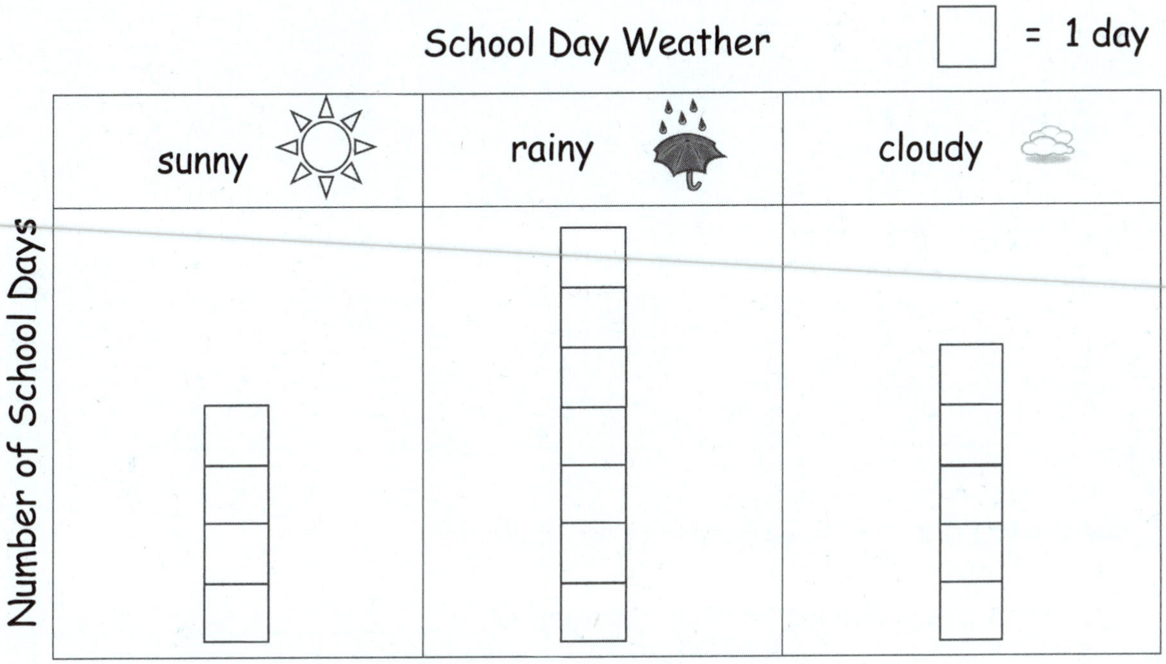

School Day Weather □ = 1 day

	sunny	rainy	cloudy
Number of School Days	□□□□	□□□□□□□	□□□□□

1. How many more days were cloudy than sunny?

   _______ more day(s) were cloudy than sunny. ______________________

2. How many fewer days were cloudy than rainy?

   _______ more day(s) were cloudy than rainy. ______________________

3. How many more days were rainy than sunny?

   _______ more day(s) were rainy than sunny. ______________________

4. How many total days did the class keep track of the weather?

   The class kept track of a total of _______ days. ______________________

5. If the next 3 school days are sunny, how many of the school days will be sunny in all?

   _______ days will be sunny. ______________________

Use the graph to answer the questions. Fill in the blank, and write a number sentence that helps you solve the problem.

6. How many fewer students chose bananas than apples?

_______ fewer students chose bananas than apples. ____________________

7. How many more students chose bananas than grapes?

_______ more students chose bananas than grapes. ____________________

8. How many fewer students chose grapes than apples?

_______ fewer students chose grapes than apples. ____________________

9. Some more students answered about their favorite fruits. If the new total number of students who answered is 20, how many more students answered?

_______ more students answered the question. ____________________

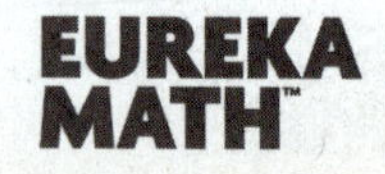

Name ___________________________ Date _______________

Use the graph to answer the questions.

Animals on Lily's Farm ☐ = 1 animal

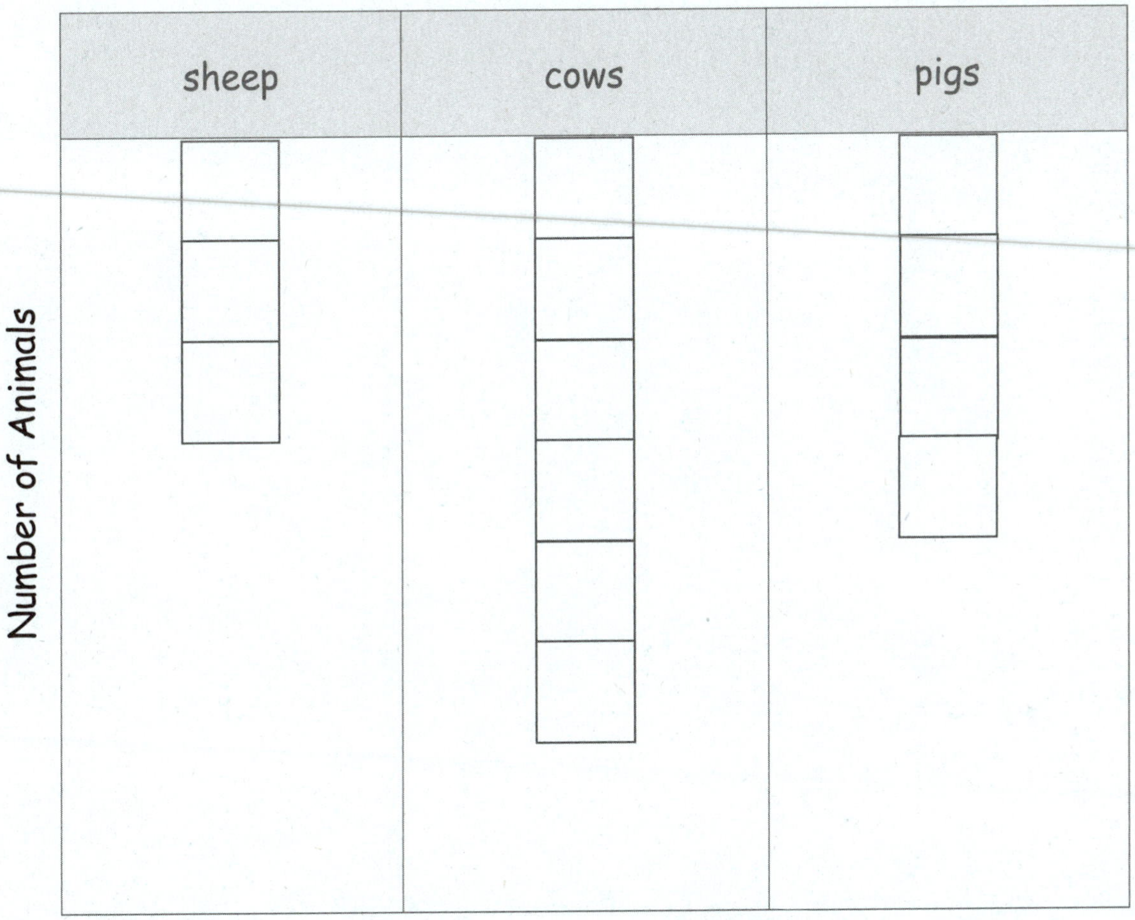

Number of Animals

sheep	cows	pigs
☐☐☐	☐☐☐☐☐☐	☐☐☐☐

1. How many animals are on Lily's farm in all? ___________ animals

2. How many fewer sheep than pigs are on Lily's farm? ___________ fewer sheep

3. How many more cows are on Lily's farm than sheep? ___________ more cows

Credits

Great Minds® has made every effort to obtain permission for the reprinting of all copyrighted material. If any owner of copyrighted material is not acknowledged herein, please contact Great Minds for proper acknowledgment in all future editions and reprints of this module.